ADHESION 8

This volume is based on papers presented at the 21st annual conference on Adhesion and Adhesives held at The City University, London

Previous conferences have been published under the titles of
Adhesion 1–7

ADHESION 8

Edited by

K. W. ALLEN

*Adhesion Science Group, Department of Chemistry,
The City University, London, UK*

ELSEVIER APPLIED SCIENCE PUBLISHERS
LONDON and NEW YORK

ELSEVIER APPLIED SCIENCE PUBLISHERS LTD
Ripple Road, Barking, Essex, England

Sole Distributor in the USA and Canada
ELSEVIER SCIENCE PUBLISHING CO., INC.
52 Vanderbilt Avenue, New York, NY 10017, USA

British Library Cataloguing in Publication Data

Adhesion.—8
  1. Adhesion—Congresses
  541.3′453   QD506.A1

  ISBN 0-85334-252-0

WITH 55 TABLES AND 77 ILLUSTRATIONS

© APPLIED SCIENCE PUBLISHERS LTD 1984
© CONTROLLER, HMSO, LONDON 1983—Chapter 7
© UNILEVER UK CENTRAL RESOURCES LTD 1983—Chapter 11

The selection and presentation of material and the opinions expressed in this publication are the sole responsibility of the authors concerned.

Printed in Great Britain by Galliard (Printers) Ltd, Great Yarmouth

Preface

The Annual Conference on Adhesion and Adhesives from which this volume derives has now clearly 'come of age' with its twenty-first meeting. In maturity it has achieved a positive and distinctive niche as a major international event in its field of study and expertise. Speakers and audience come from far and wide each Easter to interchange news and views, and this volume now presents these to a wider audience and in a more permanent form.

May I express thanks, both personally and on behalf of the City University, to all those who help with all the various facets of the conferences and the resulting books.

K. W. ALLEN

Contents

vii

1

Chapter 1

ENERGY ABSORPTION AND STRENGTH OF ADHESIVE LAP JOINTS
UNDER IMPACT LOADING

J. A. HARRIS   and R. D. ADAMS

University of Bristol, Department of Mechanical Engineering,
Bristol, UK

1 INTRODUCTION

In many structures, the ability to withstand impact loading is of
great importance, particularly from the point of view of safety. The use
of adhesive bonding as the primary method of joining in structures such as
automobiles or railway carriages is of interest at the present time. In
such cases, not only is the ability of the joints to carry the normal quasi-
static loading of importance, but also the effects of high loading rates,
corresponding to a crash situation, must be considered. Since, in general,
the mechanical properties of adhesive materials are rate dependent, it is
by no means certain that the performance of bonded joints under 'static'
conditions is the same as that under 'impact' conditions.

In the work presented, three questions have been addressed:
(i)   How is adhesive joint strength affected by impact loading?
(ii)   Can adhesive joints absorb large amounts of energy?
(iii) Can the performance of adhesive joints be understood and predicted?

The performance of a lap joint has therefore been assessed in terms
of strength and energy absorption. Joint strength is important under static
conditions, whilst energy absorption is important under impact conditions,
though this will itself be dependent on the joint strength under impact
conditions.

The approach taken was firstly to measure the effects of impact loading
on the performance of a particular adhesive joint configuration, the single
lap joint. Secondly, the mechanical properties of the adhesives as bulk
materials were measured under conditions corresponding to both static and

impact rates of loading. Finally, by employing a non-linear, large-displacement finite element model of the joint, and including the appropriate adhesive properties, the state of stress and strain in the adhesive layer was predicted under the various loading conditions.

2 EXPERIMENTAL PROGRAMME

2.1 Joint Tests

The adhesive single lap joint test piece was used for this investigation, 25.4 mm in width, with a 12.7 mm overlap length as recommended by ASTM D 1002-72. This configuration was chosen as being representative of a real joint, whilst at the same time being relatively simple. It also has the advantage of being a commonly used quasi-static test configuration, the results from which can be readily compared with impact test results.

In order to test the lap joint specimens under impact conditions, an existing pendulum-type impact machine, used in the Izod and Charpy testing of metals, was modified to accommodate the specimen geometry. The specimen was tightly clamped into the rig at both ends. At one end, the impact loading, applied by the swinging pendulum, was transferred to the specimen through a bearing block, which constrained the direction of movement of the end of the specimen, thus ensuring consistent axial loading of the specimens. The other end of the specimen was attached to a load cell which was in turn rigidly fixed to the base of the machine. The rig was fully instrumented so that the dynamic loading applied to the joint and the end displacement of the joint could be recorded during the test. For the former a piezo-electric force link was used (Kistler type 9331A), which has a high natural frequency compared with an equivalent strain gauged element, thus minimising the effects of distortion owing to the dynamic response of the load cell. For the latter, a capacitive displacement probe (Wayne-Kerr type MD1) was used. The two dynamic signals were captured in each test by a two-channel transient recorder system (Datalab DL902). From these traces, the failure load was determined and, by integrating the load versus end displacement characteristic, the energy absorbed in the test was found; this was done automatically by transferring the stored data to the memory of a mini-computer which then carried out the computation.

The impact velocity appropriate to this type of test is somewhat ill-defined. In the frontal barrier impact testing of automobiles, 30 m.p.h. (13.4 ms^{-1}) is commonly used. However, the resultant rate of loading depends not only on the incident velocity, but also on the stiffness of the element to which the load is applied. The test specimen will have a much higher stiffness than a larger practical structure, so that a reduced velocity is more applicable. Therefore, a typical incident velocity of 1.4 ms^{-1} was used in the lap joint tests, controlled by the height from which the pendulum was released. This results in times to failure of the order of milliseconds. Typically, the major part of the deformation processes in an automotive impact at 30 m.p.h. will take place in the first 20 milliseconds of the event^1. Thus, the lap joint test corresponds to a relatively severe impact, even though the impact velocity is reduced.

The details of the adhesive and adherend materials used in the testing are as follows:

Adhesives: MY750 - A basic epoxy material cured with an anhydride hardener and accelerator at 80°C for 6 hours, then at 125°C for 5 hours.

AY103 - A plasticised epoxy material cured with an amine hardener at room temperature, followed by a post cure of 3 hours at 100°C.

ESP105 - A toughened epoxy material with an aluminium filler, cured at 160°C for 30 minutes.

CTBN - A toughened epoxy material consisting of MY750 and 15 parts per hundred by weight of CTBN synthetic rubber cured with 5 parts per hundred of piperidine at 120°C for 16 hours.

Adherends : 2L73 - A high strength clad aluminium alloy with a 0.2 per cent proof stress of 430 MPa.

BB2hh - An aluminium alloy containing 2 per cent Magnesium in its 'half hard' state with a 0.2 per cent proof stress of 220 MPa.

BB2s - As above, but in its 'soft' state resulting in a 0.2 per cent proof stress of 110 MPa.

Using the four adhesives and three adherend materials, batches of six joints of each of the twelve combinations were manufactured and tested in the impact rig and under quasi-static conditions in a screw driven test machine with a cross-head speed of 2 mm min^{-1}.

2.2 Bulk Adhesive Tests

For the analysis, the uniaxial tensile stress-strain properties of the adhesive materials were required. Bulk specimens of the materials were machined

from pre-cast blocks. In producing these blocks, it was necessary to ensure
that exotherming of the relatively large volume of material did not take place.
In order to avoid this problem for the ESP105 material, it was found to be
necessary to reduce the cure temperature to 90°C and cure for 12 hours. Quasi-
static tests were carried out on the bulk specimens in a servo-hydraulic test
machine. Impact tests were performed in the pendulum test rig, suitably modified to apply the tensile loading on the specimens. In order to avoid premature
failure of the specimens in these tests, the surface of the gauge length was
highly polished. In the static tests, the surface was protected from the
extensometry by a layer of varnish at the points of contact. In the impact
tests, strain was measured from resistance strain gauges bonded to the specimen
surface. These inevitably produced a small stress concentration on the surface,
leading to a slight premature failure of the specimen.

3   RESULTS AND DISCUSSION

3.1 Joint Strength

Figure 1 shows typical impact loading responses measured in tests on
joints manufactured with each of the three adherend materials and ESP105
adhesive. As indicated by these traces, failure of the joints took place in
time periods in the range 1 to 5 milliseconds, this being sufficiently long
such that stress wave effects are not apparent and relatively smooth loading
of the specimen takes place. With the high strength 2L73 adherends, apart from
small dynamic effects, there is an almost linear rise in the load to failure.
With the lower strength adherends, BB2hh and BB2s, after the initial linear
rise, a lower load is attained after which the load increases much less rapidly
and failure takes place after a progressively longer time period at a progressively lower load. Examination of the fractured joints showed that, with the
lower strength adherends, more permanent deformation of the joint took place,
particularly bending which led to rotation of the overlap under load. Thus,
the observed behaviour of the various joints is attributed to the yield and
subsequent plastic deformation of the adherends.

In the impact tests, once the peak load had been reached, fracture of
the joints took place very rapidly as indicated by the rapid drop in the load.
By taking the peak load to be the strength of the joint under impact conditions,
this could be compared with the quasi-static strength measured. Figures 2,3
and 4 show the joint strengths under the two loading conditions for each of the
four adhesives with 2L73, BB2hh and BB2s adherends respectively. For each
adherend material, the trend of these results is the same. With CTBN, the

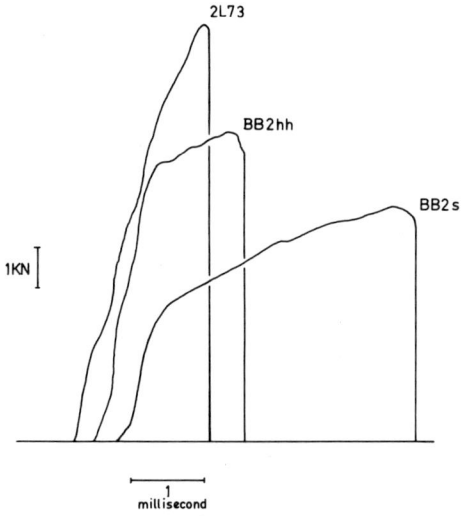

FIG.1 TYPICAL IMPACT LOADING RESPONSES FOR
VARIOUS ESP105 JOINTS

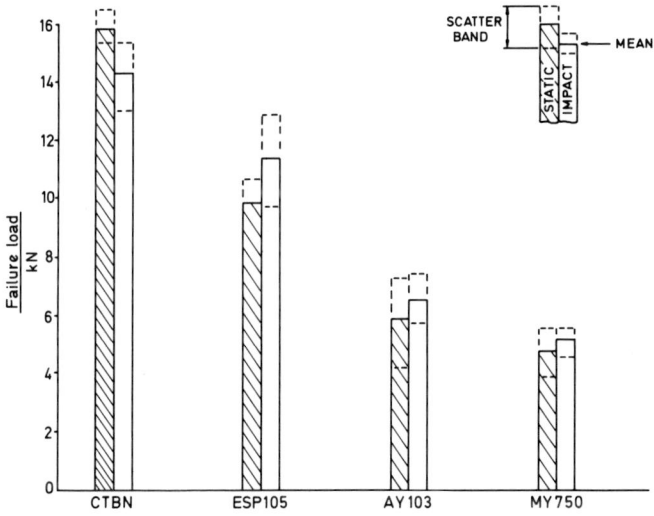

FIG. 2 STATIC AND IMPACT JOINT STRENGTHS WITH 2L73 ADHERENDS

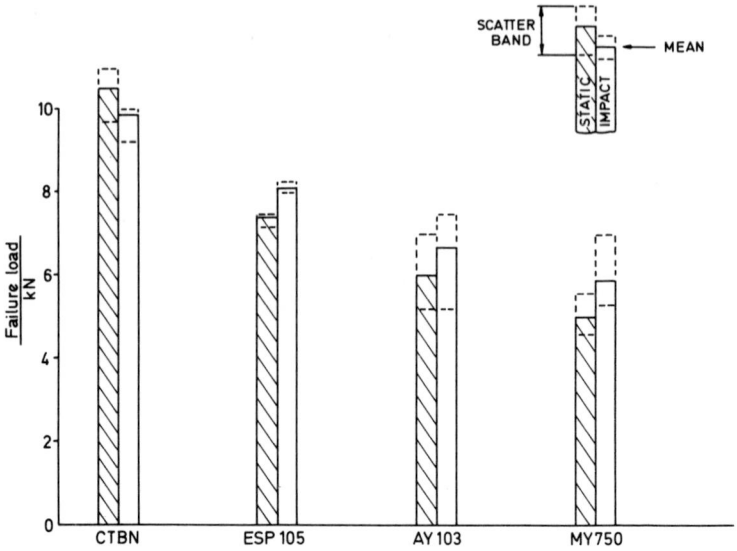

FIG. 3 STATIC AND IMPACT JOINT STRENGTHS WITH BB2hh ADHERENDS

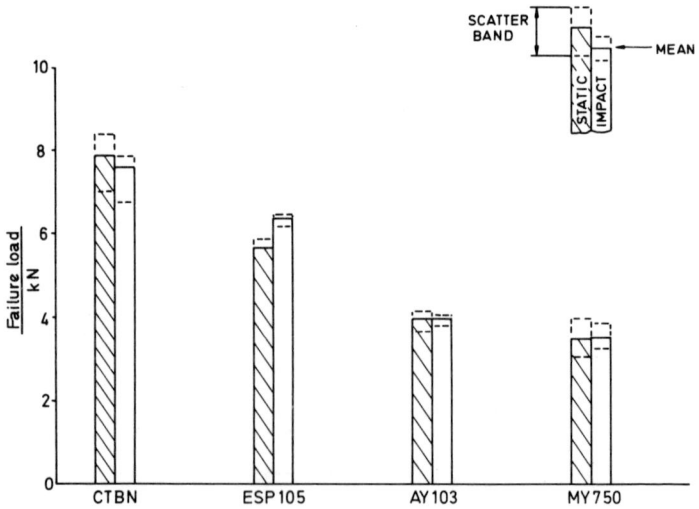

FIG. 4 STATIC AND IMPACT JOINT STRENGTHS WITH BB2s ADHERENDS

joint strength is reduced by the impact loading, whilst, for the other
adhesives, joint strength is actually increased under impact, though for the
untoughened adhesives, MY750 and AY103, the increase is not always significant
compared with the scatter of the results. For every combination of adhesive
and adherend, the effect of impact loading on joint strength is not significantly large.

In fig. 5, the results for joint strength from the impact tests have
been replotted as a function of the measured tensile yield strength of the
adherend material. For the joints with the toughened adhesives CTBN and ESP105,
the joint strength is reduced by using lower yield strength adherends. With
the other, untoughened, adhesives there is a yield stress for the adherend that
will give the highest joint strength; above this, the strength of the joint
is reduced by a small amount and below this the joint strength rapidly decreases.
Also shown in fig. 5 are lines indicating the applied load at which initial
and gross yielding of the adherend takes place. 'Initial yield' is defined as
the load at which the combination of tension and bending stresses in the
adherend at the edge of the overlap reach the yield stress of the adherend.
'Gross yield' is defined as the load at which the complete cross section of the
adherend attains the yield stress assuming only tensile loading. At loads
above the gross yield line, large scale plastic deformation of the adherends
takes place and this is associated with the rapid reduction in joint strength
with reduced adherend strength. Goland and Reissner[2], in their analysis of
the single lap joint, showed that, as the applied load is increased, because
the bending of the adherends results in rotation of the overlap, the peaks in
the stress concentrations at the edges of the adhesive layer are reduced. This
was related to a reduction of their defined bending moment factor K. The process
of adherend yielding and subsequent plastic deforming will promote further joint
rotation and may be thought of as resulting in a further reduction in K and
hence a reduction in the peak adhesive stresses. In this way, the strength of
the MY750 and AY103 joints was slightly higher with BB2hh adherends than with
2L73 adherends. However, this effect is only seen in the initial stages of
adherend plastic deformation; with gross yielding, joint strength drops off,
so that some other failure mechanism must be taking place in these cases.

From the results in fig. 5, on a strength basis, the 2L73 adherend and
CTBN adhesive combination would be chosen as giving the best joint performance.
The energy absorption characteristics of the various joints will now be considered. The energy absorbed by the joints in the impact tests has been plotted
as a function of joint strength in fig. 6. For the joints with 2L73 adherends

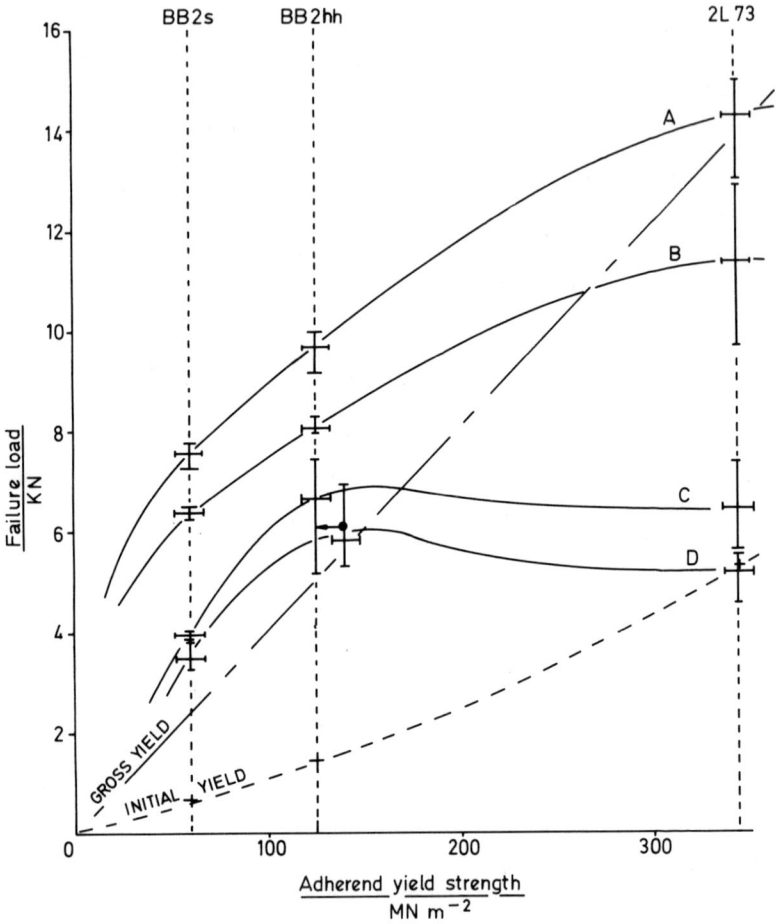

FIG. 5 VARIATION OF JOINT STRENGTH UNDER IMPACT CONDITIONS
WITH ADHEREND YIELD STRENGTH

A-CTBN, B-ESP105, C-AY103, D-MY750.

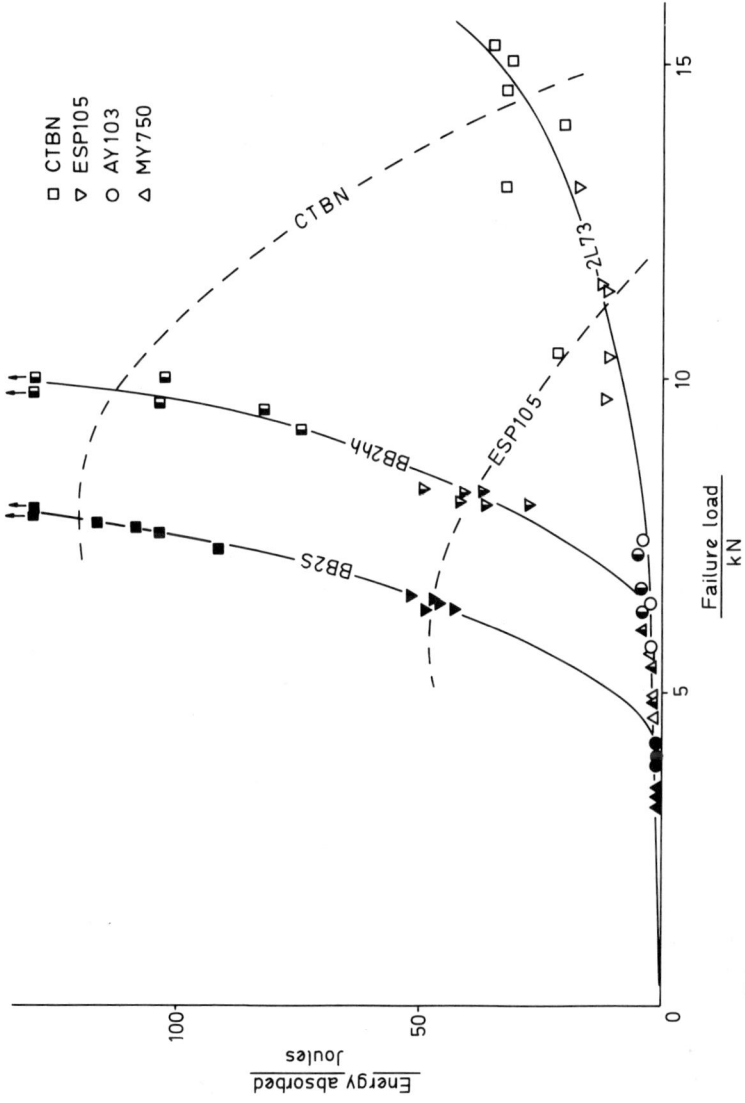

FIG.6 VARIATION OF IMPACT ENERGY ABSORBED WITH JOINT STRENGTH. FULL SYMBOLS, BB2s ADHERENDS ; HALF SYMBOLS, BB2hh ADHERENDS; EMPTY SYMBOLS, 2L73 ADHERENDS

even though significantly higher strengths are achieved with the toughened
adhesives, energy absorption is limited. However, with the lower strength
adherends, once the failure load is above a certain value, large amounts of
energy may be absorbed by the joint. It should be made clear that this energy
comes not from the fracture or yield processes in the adhesive, but from plastic
deformation in the adherends, prior to failure. Thus, even though joint
strength is decreased, a significant increase in *energy absorption* may be
brought about by using the lower strength adherends with the toughened
adhesives. Thus, on an energy absorption basis, one of the lower strength
adherends with the CTBN or ESP105 adhesive would be chosen to give the best
performance. Neither of the untoughened adhesives produces joints strong
enough for the adherend plastic deformation to take place, so it may be concluded that, under impact conditions, good performance can be achieved by
adhesive joints if sufficient joint strength is maintained such that large
scale plastic deformation can take place in the adherend materials.

3.2 Adhesive Properties

The measured uniaxial tensile stress-strain curves are shown in fig. 7
for each of the adhesives at rates corresponding to both static and impact
conditions. The effect of increased rate of loading is to increase the initial
modulus and ultimate stress and to decrease the ultimate strain in each case.
However, each adhesive shows a different degree of rate dependency which is
reflected in the relative magnitude of the changes that occur. Comparison of
these properties with the joint strength that they each produce indicates that
higher joint strength is related to increased adhesive ductility, this being
defined as the ratio of ultimate plastic to the elastic strain at failure.
This has also been shown by Hart-Smith[3], who considered the adhesive properties
in pure shear and the shear deformations of the adhesive in lap joints. The
ESP105 adhesive does not at first sight appear to fit into this pattern, because
of the relatively low ultimate tensile strains. However, its initial modulus
is approximately twice that of the other materials, because of the high proportion of filler contained in the material. Thus, even though the ultimate
strain is less than for AY103, the ratio of the plastic to elastic component
is higher i.e. higher ductility.

4 FINITE ELEMENT ANALYSIS

The analysis of the single lap joint is a non-linear problem due to the
large displacement rotation of the overlap that takes place under load. By
employing an incremental loading procedure, the finite element programme used

was able to include the large displacement effects. In the same way, the elastic-plastic properties of the adherends and the adhesive could also be included, being derived from uniaxial tensile stress-strain curves. Both the static and impact properties of the adhesives could be included by using the appropriate curve from fig. 7. Further details of the non-linear finite element programme are given in ref. 4. One further advantage of using the finite element method was that the geometry of the adhesive spew fillet, formed when excess adhesive is squeezed out of the joint during manufacture, could be accommodated.

The finite element analysis is considered here in two parts. Firstly, the effects of adherend yielding and plastic deformation are considered on the adhesive stresses, with the adhesive as a linear elastic material. Secondly, the adhesive non-linear properties are included and, by applying appropriate failure criteria, joint strength under the various conditions with various adherends, is predicted.

4.1 Adherend Plasticity

As an example of the analysis of the effects of adherend yielding on the adhesive stresses in the single lap joint, the results for a joint with the low strength BB2s adherends and an elastic adhesive will be considered. In fig. 8, normalised distributions of the maximum principal adhesive stress through the adhesive layer are shown at various levels of applied load. Initial yielding of the adherend is at a load of approximately 1.3 kN, so that at 0.01 kN no yielding has taken place and the peak adhesive stress is at point A. At 3 kN, after some adherend plastic deformation, the peak at A is reduced as a result of the rotation of the overlap under load (a process enhanced by the adherend plastic deformation). However, the peak at B is increased due to the influence of the adherend deformation in the region around the edge of the adhesive spew. After loading to 6 kN, the peak at B is increased further while the peak at A is further reduced. Thus, although the distribution of stress becomes more even through the overlap due to the enhanced joint rotation, adherend plastic deformation produces a local peak in the adhesive stresses at the edge of the adhesive spew. There are therefore two possible sites for the initiation of failure leading to the two types of fracture illustrated in fig. 9. The type I failure initiating in the region of point A was first recognised by Adams and Peppiatt[5]. However, when there is sufficient plastic deformation of the adherends, the type II failure initiating from the region of point B will occur. Thus, with the high strength 2L73 adherends, type I failure occurs

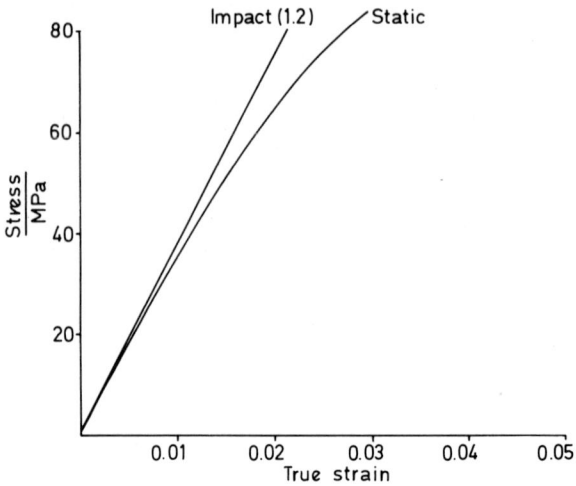

(a)

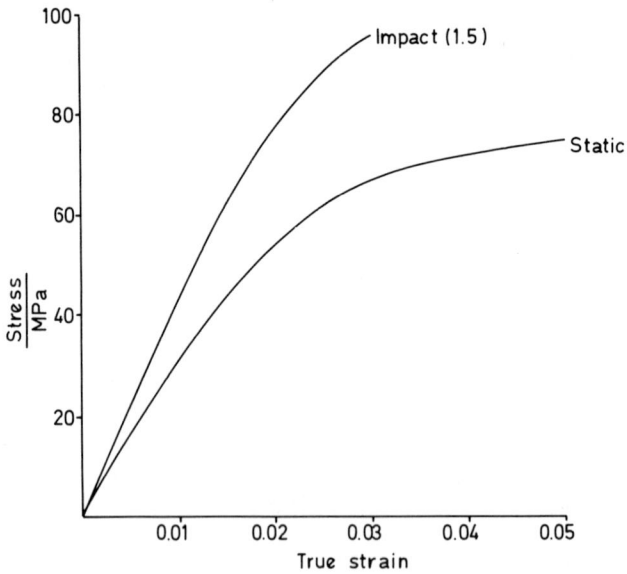

(b)

FIG. 7   UNIAXIAL TENSILE STRESS – STRAIN PROPERTIES FOR ADHESIVES
UNDER STATIC AND IMPACT CONDITIONS ; (a)MY750,(b)AY103
(c)ESP105,(d)CTBN (Approximate times to fracture in milliseconds
indicated in brackets)

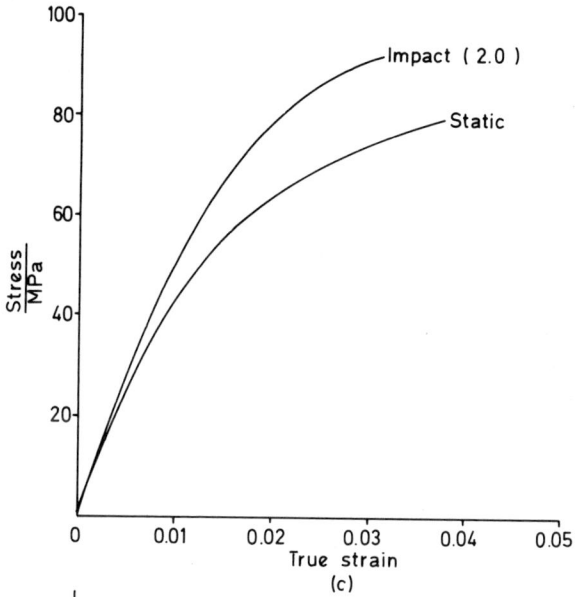

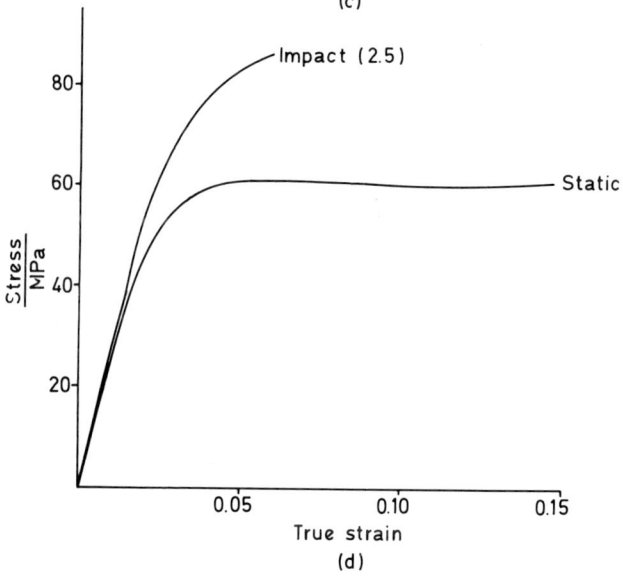

FIG.7 (cont)

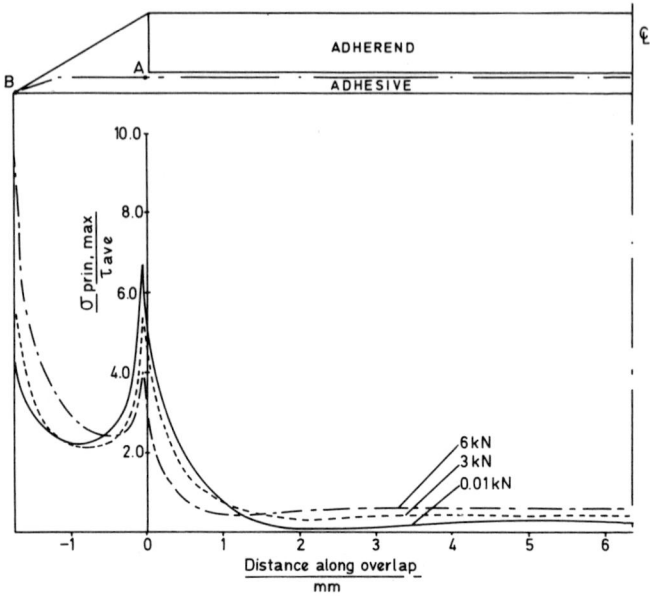

FIG.8 NORMALISED MAXIMUM PRINCIPLE STRESS DISTRIBUTION ALONG THE
ADHESIVE LAYER AT VARIOUS APPLIED LOADS, WITH BB2s ADHERENDS

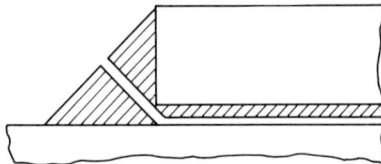

Type I mode of fracture

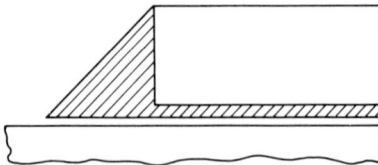

Type II mode of fracture

FIG.9   MODES OF FRACTURE IN LAP JOINTS

for all the adhesives. For the toughened adhesives, ESP105 and CTBN, when
the lower strength adherends are used, adherend yielding causes type II failure
at a lower load. For the MY750 and AY103 adherends, when BB2hh adherends are
used, some plastic deformation takes place, but not sufficient to promote type
II failure. As the analysis showed, adherend yielding leads to a reduction in
the stress concentration at point A, which explains the slight increase in
joint strength compared with 2L73 adherends. However, with BB2s adherends,
again type II failure occurs and joint strength is lower.

4.2 Adhesive Plasticity

By including the adhesive's elastic-plastic properties, the stresses and
strains in the adhesive layer are predicted by the analysis. For the states
where relatively little adhesive plastic deformation has taken place, the stress
and strain distributions are similar in form to those in fig. 8, depending on
the degree of plastic deformation in the adherends. Adhesive yielding will take
place initially in the regions of maximum stress, leading to a redistribution
of the stresses in the adhesive layer, and the more plastic deformation that
takes place, then the more even does the stress distribution become. However,
although the adhesive stresses become more uniformly distributed, the adhesive
strains become more concentrated at the points which were previously the maximum stress points. Hence peaks in the maximum adhesive principal strain distributions develop at A or B again, depending on the adherend deformation.

4.3 Strength Prediction

In order to predict joint strength from the predicted peak adhesive
stresses or strains, a failure criterion is required. In table 1, the strength
of the joints under static loading with 2L73 adherends is predicted using both
a critical maximum principal stress and a critical maximum principal strain,
equated to the ultimate values measured in the bulk adhesive uniaxial tensile
tests. By comparison with the measured strengths it would appear that, for the
untoughened adhesives, a stress criterion is more appropriate, whilst for the
toughened materials, a strain criterion fits best. This result may be what is
intuitively expected, although it may be an oversimplification to compare the
critical conditions in a uniaxial loading case with those of a triaxial loading
case on the adhesive in the critical regions in the joint.

Using the failure criterion appropriate to the particular adhesive, the
effect of adherend yielding on joint strength can now be predicted. The results
are illustrated in fig. 10 for the static joint tests. Inherent in these predictions are also the predicted modes of failure as discussed in section 4.1,

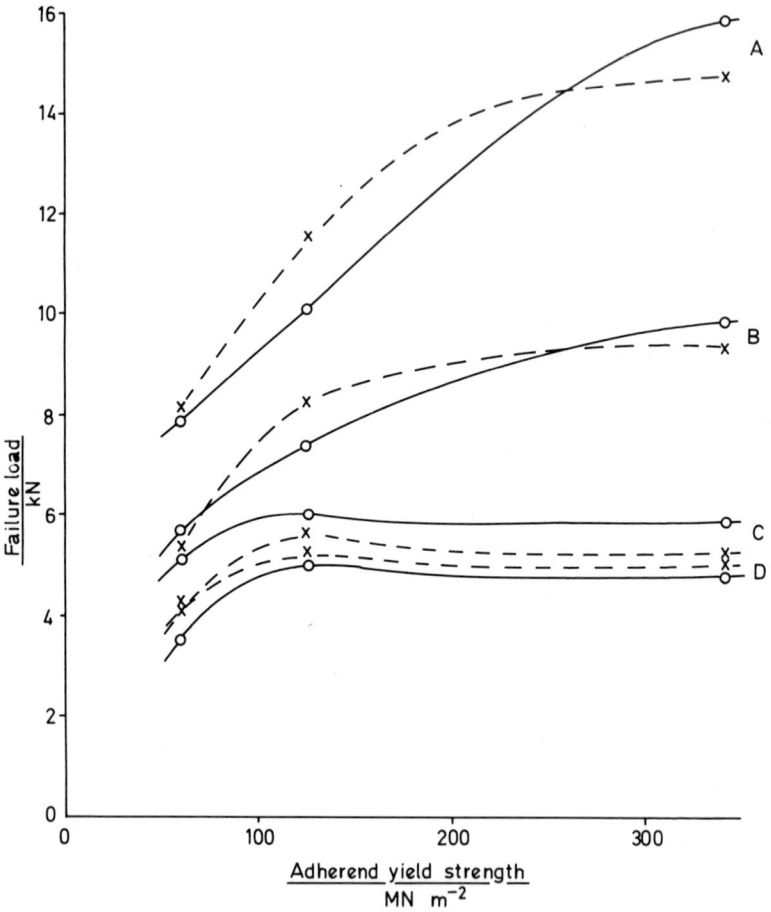

FIG.10 VARIATION OF JOINT STRENGTH WITH ADHEREND YIELD
STRENGTH

————o—— experimental   ——x——— finite element predictions

A—CTBN, B—ESP105, C—AY103, D—MY750

which together give a reasonably accurate prediction of the various joint strengths.

Table 1:

Predicted and experimental joint strengths
with 2L73 adherends under static loading

ADHESIVE	PREDICTED STRENGTH kN		EXPERIMENTAL MEAN STRENGTH kN
	STRESS CRITERION	STRAIN CRITERION	
MY750	5.05	7.2	4.8
AY103	5.3	9.3	5.9
ESP105	6.0	8.85	9.9
CTBN	4.3	14.7	15.9

Table 2:

Predicted and experimental joint strengths with
2L73 adherends under static and impact loading

ADHESIVE	PREDICTED STRENGTH kN		EXPERIMENTAL MEAN STRENGTH kN	
	STATIC	IMPACT	STATIC	IMPACT
MY750	5.05	4.5	4.8	5.2
AY103	5.3	5.9	5.9	6.5
ESP105	8.85	9.2	9.9	11.4
CTBN	14.7	10.3	15.9	14.3

Finally, again by using the appropriate failure criterion, the effect of impact loading on joint strength can be predicted for each adhesive. In table 2, the predictions of the static and impact strengths of the joints with 2L73 adherends are compared with the measured strengths. Generally, the predictions for impact strength are not so reliable, because of the difficulty in the measurement of the ultimate values of the stress and strain of the bulk material, leading to predictions that are somewhat low. However, apart from

the MY750 joints, the effect of impact loading in increasing or decreasing
joint strength, is predicted correctly.

5 CONCLUSIONS

It has been demonstrated for the particular case of an adhesive single
lap joint that impact loading does not have a significant effect on joint
strength. This has been shown for a wide range of adhesive properties.

It has also been shown that adhesive joints may absorb large amounts
of energy prior to failure. However, because adhesives are relatively brittle
materials and are present in such small volumes in a joint, the energy cannot
come from the deformation and fracture of the adhesive, but rather it must be
absorbed in the plastic deformation of the adherend materials prior to failure.
The strongest joint may not necessarily absorb the most energy, so the yield
properties of the adherends are an important consideration. By using a lower
strength adherend material, energy absorption can be significantly increased
as the joint is able to withstand large amounts of plastic deformation in the
adherends prior to failure.

By the use of non-linear finite element techniques, the various mechanisms
affecting joint strength in the lap joint tests may be understood. The onset of
adherend plastic deformation at first leads to a reduction in the peak adhesive
stress concentration, as the rotation of the overlap is enhanced. However,
significant plastic deformation results in a localised concentration of stress
at the edge of the adhesive spew leading to reduced joint strength.

Both the adhesive and the adherend elastic-plastic properties should be
included in the analysis, and, by using failure criteria applicable to a
particular adhesive, both the mode of failure and joint strength may be predicted, leading to a reasonable estimate of the effect of adherend yielding
and impact loading on the strength of single lap joints.

REFERENCES

1. Sadeghi, M.M. and Suthurst, G.D. 'Test aided computer prediction of
   passenger car side impact', ISATA Conf. 1982.

2. Goland, M. and Reissner, E. 'Stresses in cemented joints', Trans. ASME,
   J. Appl. Mech., A, 66, 17 - 27, 1944.

3. Hart-Smith, L.J. 'Stress Analysis: A Continuum Mechanics Approach',
   Developments in Adhesives - 2, 1 - 44, Applied Science, 1981.

4. Crocombe, A.D. 'The non-linear stress and failure analysis of adhesive
   tests', Ph.D. Thesis, University of Bristol, 1981.

5. Adams, R.D. and Peppiatt, N.A. 'Stress Analysis of adhesive bonded lap joints'
   J. Strain Anal. 9, 185 - 196, 1974.

Chapter 2

EFFECT OF ADHESIVE COMPOSITION ON THE PEELING BEHAVIOUR OF ADHESIVE TAPES

D W AUBREY

London School of Polymer Technology
Polytechnic of North London
Holloway Road
London   N7 8DB

1   INTRODUCTION

Many studies have been devoted to an understanding of the peeling behaviour of adhesively-bonded tapes, and among the factors often investigated are peeling angle, rate and temperature of peeling, thickness and composition of both adhesive and flexible member, and substrate surface and rigidity. This present discussion is concerned with the rationalisation of formulation changes in the adhesive, and describes some results obtained for tape joints which differ only in adhesive composition, other factors being held constant. Although the work is primarily of interest in the field of pressure-sensitive tapes, the results, being concerned exclusively with the de-bonding process, have relevance to the formulation of rubbery adhesives in general.

2   JOINT SYSTEM

The joint system under consideration comprises a thin, flexible (but essentially inextensible) backing film of polyester, bonded via a polymeric adhesive (which may be regarded as a soft rubber) to a smooth rigid substrate (in this case glass). Peeling is maintained at 90° angle and the dimensions of the joint will be unchanged (fig. 1). In considering the effect of changes in the adhesive composition on the peeling characteristics of such a joint, it is helpful to consider the response of an approximate mechanical model (fig. 1).

In the assumed mechanical analogue, which is intended only as a crude qualitative approximation to the properties of the joint, the film backing is shown as an ordinary elastic spring of very high modulus so that, provided its yield point is not exceeded (ie. provided that it does not show ductile deformation), it does not materially contribute to the work of peeling.

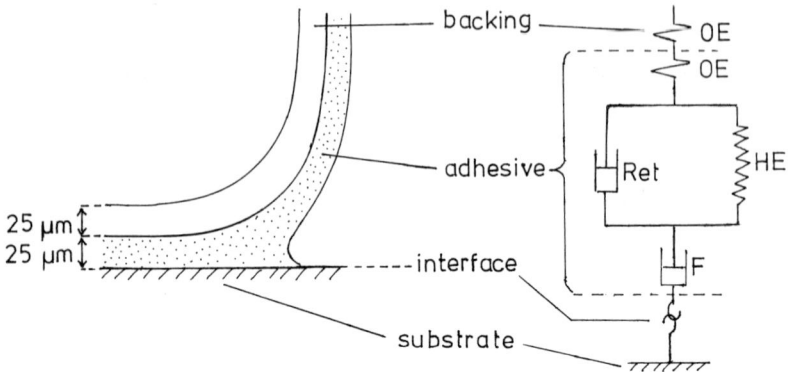

Figure 1 :   Peeling joint and assumed mechanical analogue.

The adhesive is shown as a Burger's model in which all four components, corresponding to different molecular processes, may contribute to the work of peeling
to an extent depending on rate and temperature. The glass substrate is assumed
completely rigid and undeformable throughout. The adhesive-glass interface
is represented by a hook, and separation occurs here when a critical stress,
independent of rate and temperature, is reached. In order to reach this
critical stress, a considerable amount of deformation may have to be applied
to the upper part of the model. Under normal conditions, by far the larger
part of the overall work of peeling is the work of such deformation, being
dissipated by viscous deformation in the 'retardation' and 'flow' dashpots.
The work done in breaking the hook (ie. in separating unit area of interface)
is much smaller and may in this case be equated with the reversible thermodynamic work of adhesion, W_a, due to Van der Waals interactions only. This
interfacial work, although small, is clearly important in determining the
overall work necessary to effect separation. The overall Work of Peeling,
W_p, may then be expressed in the form:

$$W_p = W_a + W_a \emptyset$$

where Ø is a 'dissipation factor' depending
on the rate, temperature and extent of
adhesive deformation

or, if the interfacial work is negligible:

$$W_p = W_a \emptyset \qquad \qquad \dots \dots \dots \dots \dots \dots \dots (1)$$

It is possible to clarify the significance of the various components in the above mechanical model by keeping constant two of the parameters controlling $\emptyset$ and varying the third.

In our joint (fig 1) the extent of deformation, or strain, is kept essentially constant by maintaining a constant adhesive thickness. If we also keep temperature constant, and plot work of peeling ($\propto$ peel force P) against rate, it is possible to resolve qualitatively the contributions of the various components in the Burger's model of fig. 1.

3 EFFECT OF PULLING RATE

The apparatus and method used for peeling tapes at 90° angle from glass have been described in an earlier contribution[1] to this series. Also described was the application of time-temperature superposition in order to obtain a 'master-curve' of peel force over a large range of rates at a standard temperature. These methods were used [1,2] in obtaining the force-rate relationship of fig. 2 by using an uncrosslinked amorphous elastomer (poly-n-butyl acrylate) as the adhesive.

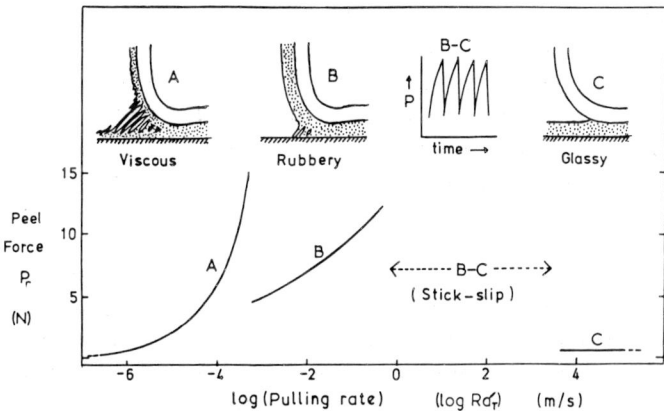

Figure 2 : Dependence of peel force on (log) pulling rate at 296 K for polyester backing/polybutyl acrylate adhesive/glass substrate. Tape is 2.54 cm wide, and other dimensions as shown in figure 1. Drawings illustrate the adhesive behaviour in each region. (From ref. 2)

In the force - rate 'master-curve' (fig 2) there are four distinct modes of peeling:

Region A: At these low rates, separation is 'cohesive', involving much filamentation or 'legging' within the adhesive, and force is highly rate - dependent. The predominant adhesive response is viscous flow (ie. deformation of the 'flow' dashpot in figure 1). There is insufficient stress at the interface for separation to occur.

Region B: Separation is now 'adhesive' at the polymer - glass inter-face, the adhesive shows little filamentation, and force is moderately rate - dependent. The predominant adhesive response is rubbery (retarded high elastic). This involves stretching the Voigt element (high elastic spring and retardation dashpot in parallel) of figure 1. This kind of peeling, in which the adhesive shows 'rubbery' behaviour, is the one of greatest inter-est for pressure-sensitive tapes and for rubber-bonded joints generally.

The increase in force with rate in regions A and B may be thought of in terms of increasing energy dissipation arising from increasing stiffness of the 'flow' and 'retardation' dashpots, respectively.

Region C: Separation occurs at the adhesive - backing interface, there is no filamentation, force is very low and rate-independent, and the adhesive surface after peeling is smooth and shiny. This is believed to involve a glassy response in the adhesive, ie. deformation only of the ordinary elastic springs representing the backing and the adhesive in figure 1, the stiffness of both dashpots being too high to allow any viscoelastic deforma-tion.

Region B-C: In this so-called 'stick-slip' or oscillatory mode of peeling, the force oscillates regularly between fairly well-defined limits (a typical testing machine trace being shown in the inset diagram of figure 2). The response of the adhesive is believed to alternate between rubbery (region B) and glassy (region C), and is associated with the storage and dissipation of elastic energy in the stretched free (already peeled) part of the tape. A second, very narrow, region of stick-slip behaviour (region

A-B) is sometimes observed between regions A and B[3].

It is emphasised that the above characteristics are not unique to
this particular joint system. Master-curves showing the same basic features
have been obtained using other materials and other peeling conditions, eg.
in peeling alkyl acrylate copolymer adhesives at 180° angle from glass and
other substrates[4], in T-peeling of cloth-supported SBR adhesives from polyester
film[5], and in peeling polyimide film-supported polystyrene adhesive from chromium metal[6].

Now, there are three principal ways in which a change is the nature
of the adhesive polymer may modify the peeling characteristics as illustrated
in figure 2.

First, the adhesive may be changed in its chemical composition (the
nature of the chemical groups present). Such changes affect segmental mobility
and are conveniently reflected in changes in the adhesive glass transition
temperature, T_g. This is equivalent to changing the viscosities of the liquids
in the dashpots of figure 1, and it changes the rate (or temperature) at
which transition from one viscoelastic state to another occurs. Thus, this
effect would be expected to change the rates at which the transitions between
regions A, B and C occur (fig. 2).

Secondly, the average molecular weight of the adhesive may be changed,
to affect mainly the flow behaviour, ie. it affects mainly the viscosity of
the flow dashpot of figure 1, and only to a small extent (except at very
low molecular weights) that of the retardation dashpot. Thus, an effect
is expected on the level of peel force in region A and possibly on the rate
at which it changes to region B (fig. 2).

Thirdly, a change in the nature of the adhesive may change the Work
of Adhesion, which depends on the interfacial free energy, ie on the strength
of the hook in figure 1. This is purely an interfacial effect, not necessarily
connected with any change in the adhesive bulk, and is therefore independent
of rate and temperature. The expected effect would be a vertical shift of
the force-rate curve in regions of interfacial separation (eg. region B).
The same kind of effect would, of course, be observed with changes in the
chemical nature of the substrate surface.

Formulation changes to an adhesive will normally affect all three

of the above to varying degrees, resulting in changes in the peeling characteristics which may be superimposed and indistinguishable from each other.
Nevertheless, we will examine the effects of some systematic changes to the
adhesive formulation, to see to what extent these individual effects may
be resolved.

4 EFFECT OF CHEMICAL COMPOSITION OF ADHESIVE

If the polybutyl acrylate in the joint of figure 2 is replaced by
polyvinyl acetate of similar molecular weight, the force-rate plot shown
in figure 3 is obtained[2]. The overall pattern of peeling behaviour is clearly
very similar to that of figure 2, except that the whole surve has been displaced some 12 decades to the left along the abscissa. This lateral shift
reflects the large difference in T_g between the two polymers, and means that
their peeling characteristics are similar, but over a different range of
rates (or temperatures). Although there are some slight differences attribut-
able to the second and third effects referred to above, it is clear that
the most dramatic change is the lateral shift due to the change in bulk properties of the adhesive in altering its basic chemical composition.

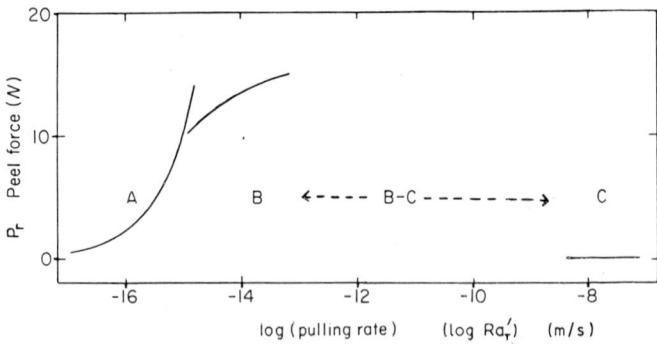

Figure 3:   Force - rate plot for the same joint system as that
             of figure 2, except that polyvinyl acetate has replaced
             the polybutyl acrylate adhesive.

It will be apparent that polyvinyl acetate itself cannot be regarded
as a good adhesive for bonding tapes, since it will exhibit glassy response

at typical pulling rates (say, 1 cm/sec, ie log Ra_T=-2 m/s). Polybutyl acrylate shows rubbery response at this rate, but it is clear from fig. 2 that a slight shift to the left might be advantageous in that it would produce a higher peel force at this rate. Such a slight shift can be accomplished by copolymerisation with minor amounts of other monomers such as vinyl acetate or acrylic acid, and is considered later.

The amount of shifting of region A (cohesive separation) between figs. 2 and 3 may be predicted accurately from their glass transition temperatures T_g using the WLF equation. Shifting of region B is somewhat less certain since there is a slight upwards shift of this region in fig 3 due to the higher interfacial attractions of polyvinyl acetate to glass. Similar results were obtained in peeling paper-supported acrylate copolymer adhesives from glass, using adhesives of various T_g's [7], where again prediction of the shifting was accurate only in the region of cohesive separation.

5 EFFECT OF ADHESIVE MOLECULAR WEIGHT, AND CROSSLINKING

A change only in molecular weight, MW, of the adhesive has very little effect on the location of the whole curve on the abscissa, unless MW is extremely low. This would be expected in view of the well-known slight effect of MW on T_g, except at very low MWs.

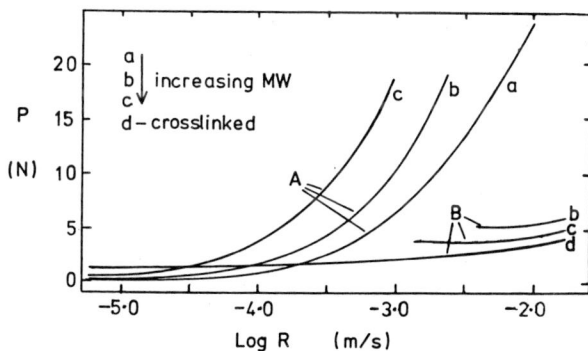

Figure 4: Force - rate plot for the same joint system as that of figure 2, with variation in MW and crosslinking of the polybutyl acrylate.

The main effect of MW is seen in the flow region, ie, at the left-hand end of the force-rate master-curve of figures 2 and 3. The effect may be illustrated by changing the MW of polybutyl acrylate, as shown in figure 4.

Figure 4 shows only the left-hand end of the master-curve and gives results of tests at room temperature (296K). Complete master-curves, analogous to those of figs. 2 and 3, may be obtained from these systems by using time-temperature superposition; the curves so obtained gradually become more co-incident towards the right and differ only slightly in position in the stick-slip (B-C) and glassy (C) regions.

The results shown in fig 4 (re-drawn from refs 2 and 3) show that, as MW is increased, peeling with rubbery response (region B) is extended towards lower pulling rates, at the expense of region A. When the MW becomes infinite, ie. a network is formed by crosslinking, flow is no longer possible and peeling region A disappears altogether. These observations are analogous to the well-known extension of the rubber 'plateau' with molecular weight in other viscoelastic master-curves such as dynamic shear modulus versus frequency or temperature. A small part of the lateral displacements of the curves in figure 4 is due to change in T_g with molecular weight (see section (4), above), but these may be neglected compared with the effect of MW on flow behaviour. Although these results do not show the effect of changing the degree of crosslinking, it may be noted that, in studies of SBR peeled from a fluoropolymer surface, there was no effect other than a lateral shift due to change of T_g with crosslinking[8].

Increasing the MW or crosslinking an adhesive polymer may thus be important because it broadens the range of conditions over which the more desirable rubbery mode of peeling is observed. It is probably more important, however, in improving the shear strength under steady load (an effect outside the scope of the present discussion); tapes for high temperature or long-term load-bearing appllications will be cross-linked, whereas labels, decorative sheets and other non-load-bearing products are seldom crosslinked.

It will be apparent from figure 4 that, for tapes with crosslinked adhesives, peel force becomes very low at low rates (or high temperatures), usually a disadvantageous effect in practice. As rates are reduced the peeling force theoretically approaches that required for interfacial separation, since viscoelastic dissipation processes should disappear (the model of figure 1 then approximates to a high elastic spring only). Values of peeling energy for SBR from various substrates showed, at very low rates, quite good agreement with the thermodynamic work of adhesion, provided that only secondary intermolecular forces operated at the interface[9].

6 EFFECT OF CHANGE IN BOTH CHEMICAL COMPOSITION AND MOLECULAR WEIGHT OF
   THE ADHESIVE

Adhesives are normally formulated by the mixing of different materials, and formulation changes will usually involve modification of both T_g and molecular weight. As long as there is negligible change in interfacial free energy, therefore, we may expect to see both effects (4) and (5) above in the modified peel force - rate master-curve. Systems in which negligible change in interfacial energy might be expected are the incorporation of ester plasticisers into ester polymers (such as polyvinyl acetate or vinyl acetate/ alkyl acrylate copolymers), and the incorporation of alicyclic tackifier resins into hydrocarbon rubbers (such as natural rubber).

Some results[10] showing the effects of progressive addition of a terpene tackifier resin ('Piccolyte S115') to natural rubber are shown in figure 5.

In figure 5, the curves for NR only and for a 75/25 rubber/resin blend could not be extended into the stick-slip region B-C because of fracture of the polyester backing. Nevertheless, the anticipated trends may be seen in these curves.

The most obvious effect of increasing resin concentration is to shift the master-curves in the direction of lower pulling rates. This is most easily seen by examining the positions of the rubbery region (region B) at high rates, and the glassy region (region C). This shift is associated with the rise in T_g as resin concentration increases (T_g values corresponding to curves (a), (b), (c) and (d) are 202K, 210K, 227K and 278K, respectively).

A second effect revealed in figure 5 is a progressive extension of region A to higher peel force values as resin concentration increases. This is the same effect as that discussed under section (5) above and is due to the progressive reduction in average molecular weight as the (low MW) resin concentration increases. The shift in the region A — region B transition towards higher rates, observed in section (5) above, is disguised in this present case because the whole master-curve is being displaced to the left at the same time, by the first effect referred to above. Thus, because the tackifier resin causes an increase in T_g it shifts the whole curve to the

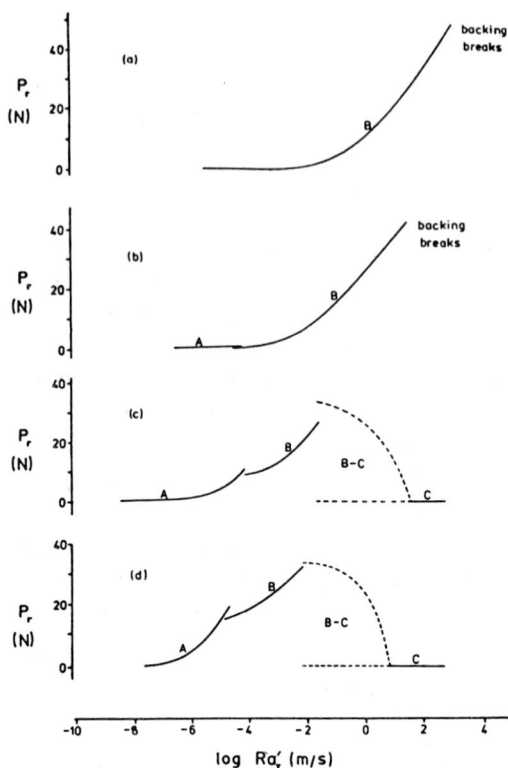

Figure 5: Peel force Pr versus (log) pulling rate, log Ra$_T$ at
296K for the same joint system as that of figure 2,
except that the polybutyl acrylate is replaced by
blends of NR and terpene resin ('Piccolyte S115')
in various proportions.

(Wt ratio NR/resin: (a) 100/0 (b) 75/25 (c) 60/40
(d) 40/60)

left, and because it reduces average MW it moves the left-hand end of it
to the right. The first shift is important in determining peel strength
at 'common' pulling rates (compare, for example, peel force for the four
compositions at 1cm/s, ie log Ra$_T$ = -2 m/s); the second shift, affecting
flow at very low rates, reflects the increase in 'tack' with resin concentration by an increased ability to make rapid contact with surfaces. The latter
effect is, however, not within the scope of this present discussion.

Analogous behaviour was shown in a study of the temperature - dependence
of peel strength of tapes with different contents of tackifier[11], where

increasing tackifier content caused the adhesive → cohesive (B → A) transition to move to lower temperatures, and adhesive regions of equivalent peel
force at low temperatures (≡ upper end of B) moved to higher temperatures.
These effects caused by changing temperature are clearly the same as those
produced by changing rate in fig 5.

It will be evident from the foregoing considerations that a tackifier
resin can influence both the 'bonding' and 'de-bonding' properties of a pressure-sensitive adhesive. It seems that, for efficient action, a tackifier
resin should have a low as possible a MW combined with as high as possible
a T_g. A highly condensed alicyclic ring structure appears to be necessary
to achieve this combination satisfactorily.

7 EFFECT OF INTERFACIAL FREE ENERGY

Where a change in adhesive composition is likely to cause an appreciable
change in interfacial free energy (and, therefore, in Work of Adhesion),
we expect to see a vertical shift in the curve for interfacial separation
(region B), superimposed on any of the above lateral shifts caused by the
change. An example of such a change is the introduction to carboxylic acid
groups, by copolymerisation, into the adhesive. This is known to produce
a dramatic increase in peel strength from hydroxylic surfaces such as glass,
an effect which is made use of commercially and is usually attributed to
hydrogen bonding across the interface. However, it will be evident from
the earlier considerations that, if the incorporation of carboxylic groups
raises the T_g, there will also be an effect on the adhesive bulk which will
cause peel force to rise (when measured at a given rate). Thus, it is not
usually easy to resolve the overall effect of a formulation change into its
individual effects on bulk viscoelastic behaviour (ie changes in T_g and MW)
and on interfacial attractions (interfacial free energy).

It is, of course, easy to vary interfacial free energy, without changing
bulk properties merely by peeling from different surfaces, and indeed this
has been extensively studied (eg refs. 4, 9, 12). However, the present discussion concerns formulation changes within the control of the adhesive manu-
facturer, and therefore considers changes to the adhesive rather than the
substrate.

Using our peeling joint, the introduction of 10% by weight of acrylic
acid into the polybutyl acrylate by copolymerisation, followed by light crosslinking (heating with 2% benzoyl peroxide) to eliminate viscous (region A)

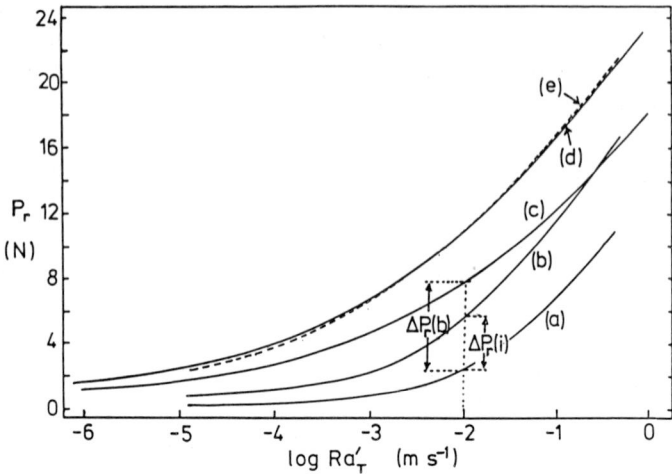

Figure 6:   Illustration of the relative contributions by inter-
             facial and bulk effects to the peeling behaviour of
             carboxylated polybutyl acrylate, using a joint system
             otherwise the same as that of figure 2.

behaviour, causes the curve of region B to change its position from curve
(a) (pure PBA) to curve (d) (carboxy-PBA), in figure 6[13]. The difference
between these two curves clearly shows a very large increase in peel force
at any chosen rate. This study was not extended into the stick-slip (B-C)
or glassy (C) regions of behaviour.

Since the curve of region B does not show precise end-points, it is
not possible to distinguish between vertical shifting (from interfacial
effects) and lateral shifting (from bulk effects) in moving from curve (a)
to curve (d). Thus, we do not know the relative importance of the interfacial
and the bulk effects in improving the peel strength at a particular rate.

In order to differentiate between these effects, the peeling behaviour
of an adhesive which had been carboxylated on the surface (but not in the
bulk) was compared with that of an adhesive which had been carboxylated in
the bulk (but not on the surface). Surface modification was carried out
by means of a very thin supercoating applied to each type of adhesive.

The surface-modified adhesives gave the force-rate curves shown in
figure 6 as curve (b) (PBA bulk with carboxy-PBA surface) and curve (c)

(carboxy-PBA bulk with PBA surface). The differences between curves (a) and (b) or between (c) and (d) should thus reflect only the interfacial effects of carboxylation (ie. should represent only a vertical displacement). The difference between curves (a) and (c) or between (b) and (d), on the other hand, reflect only bulk effects of carboxylation (ie represents only a lateral displacement). More evidence justifying these conclusions is discussed in reference 13.

The relative contributions by interfacial and bulk effects are clearly not constant as rate varies. This may be seen in the increase of ratio $\triangle P_r$(bulk) : $\triangle P_r$(interfacial) in figure 6 from about 0.3 to 1.1 over the range of rates used. Thus, the bulk effect makes the major contribution to peel strength over most of the range, but it is eventually overtaken by the interfacial effect. Since the bulk effect is independent of the nature of the interface, it will be observed during peeling from any rigid surface. Summation of the bulk and interfacial effects gave the hypothetical composite curve (e) (fig. 6), which is very similar to the experimental curve (d).

It follows from eq. 1 (section 2, above) that plots of log W_p against log Ra_T for adhesives of the same bulk but different surface compositions should be displaced vertically relative to one another by a constant interval $\triangle$ log W_a. In such plots from the results of fig. 6, the interval showed some deviation in constancy for both sets of curves [13], and corresponded to a fractional increase in W_a from surface carboxylation (ie, W_a (surface carboxy-lated/W_a (surface uncarboxylated) of between about 1.4 and 2.0. The lack of constancy of this ratio was attributed to variation in the average strain in the peeling adhesive, due to filamentation and possibly other effects. Similar plots from a study of crosslinked polybutadiene peeled from silane-treated glass surfaces[12] were parallel at low rates but not at high rates, again showing that the simple factorisation expressed by eq. 1 does not hold universally.

8 CONCLUSIONS

During the peeling of an adhesive tape from a rigid surface, the force necessary and the mode of separation are strongly dependent on the bulk visco-elastic response of the adhesive, as well as on the nature of the interface. The influence on peel behaviour of a change in adhesive formulation may be rationalised by consideration of the individual contributions by three different effects:

(i)   change in adhesive glass transition temperature;

(ii)   change in adhesive average molecular weight;

(iii) change in interfacial free energy.

REFERENCES

1. Aubrey, D.W., Chapter 12, p. 191 in "Adhesion-3", K. W. Allen (Ed), Applied Science Publishers, London 1977.

2. Sheikh, A.A., Thesis for PhD. (CNAA), London, 1973.

3. Aubrey, D.W., Welding, G.N., and Wong, T.K.M., J. Appl. Polymer Sci., 13, 2193 (1969).

4. Kaelble, D.H, J. Adhesion, 1, 102, (1969).

5. Gent, A.N., and Petrich, R. P., Proc. R. Soc. (Lond.) A310, 433 (1969).

6. Parsons, W.F., Faust, M.A., and Brady, L.E. J. Polymer Sci. (Polym. Physics) 16, 775 (1978).

7. Chan, H.K., and Howard, G. J., J. Adhesion, 9, 279, (1978).

8. Andrews, E.H., and Kinloch, A.J., J. Polymer Sci. (Polym. Physics) 11, 269 (1973)

9. Andrews, E.H., and Kinloch, A.J., Proc. R. Soc., (Lond.) A332, 385 (1973)

10. Aubrey, D.W., and Sherriff, M., J. Polymer Sci., (Polym. Chem.), 18, 2597 (1980)

11. Hata, T., J. Adhesion, 4, 161 (1972).

12. Ahagon, A., and Gent, A.N., J. Polymer Sci., (Polym. Physics), 13 1285 (1975).

13. Aubrey, D.W., and Ginosatis, S., J. Adhesion, 12, 189, (1981).

Chapter 3

ADHERENCE BETWEEN A RIGID SPHERICAL PUNCH AND AN ELASTOMERIC SOLID

M. BARQUINS

E.R. Mécanique des Surfaces du C.N.R.S.
Laboratoire Central des Ponts et Chaussées, Paris, France

SYNOPSIS

The Sneddon solution of the Boussinesq problem is used to study the adhesive contact between a rigid spherical punch and an elastomeric solid. It is shown that the introduction of the concepts of fracture mechanics allows one to study the stability of the equilibrium and to predict the kinetics of the adherence of elastomers ; the edge of contact being seen as a crack propagating in mode I in the interface. It also allows solution of the problem of the tackiness of elastomers and enables one to take into account the dwell time effect and the influence of the stiffness of testing machine on the adherence.

Experiments carried out with a glass ball in contact on a polyurethane surface, at fixed load, cyclic load, fixed displacement or cross-head velocity, verify theoretical predictions with an accuracy better than two percent.

1 INTRODUCTION

The problem of the contact area and the elastic indentation of a rigid body, schematized by a rigid sphere of radius R and a surface of elastic solid, under the action of a force P, was solved by Hertz[1], neglecting the surface energies of solids. The contact radius a_H is given by $a_H^3 = PR/K$ with $K = 4E/3(1-\nu^2)$, where E and ν are the Young's modulus and Poisson's ratio of the elastic solid, and the elastic displacement is $\delta = a_H^2/R$, i.e. twice the height of the spherical cap of radius a_H. The principal stresses and their isostatics in an elastic solid subjected to indentation by a normal load applied by means of a rigid sphere have been computed and plotted with $\nu = 0.25$ by Lawn[2] and for the

particular case of rubber-like materials (ν = 0.5) by Barquins[3]. In the latter case, on the surface all the stresses are equal inside the contact area,and
vanish outside.

But, as pointed out by Johnson et al.[4] (JKR theory), Hertz's results do not
take into account the existence of molecular attraction forces which produce infinite stresses at the edge of contact area, and tend to increase the contact
area and the elastic displacement. As shown by Johnson[5] the distribution of the
normal stresses in the contact area results from the super-imposition of hertzian
compressive stresses and of tensile stresses corresponding to a flat punch ; in
consequence, the normal stresses are tensile at the edge of the contact area and
compressive at its centre. Sneddon[6] has generalized Hertz's calculations to surfaces of any shape having axial symmetry, using Hankel's transform and Abel's in-
tegral, and it was shown in refs 7 and 8 that the proposed solution allows the
study of adhesive contacts, if an integration constant is supposed to be non-zero.
This constant is proportional to the stress intensity factor at the edge of the
contact area, so that calculations of normal stress σ_z and the discontinuity of
the displacement $[u_z]$ lead to formulae identical to those of fracture mechanics in
mode I. It would therefore seem that the contact edge may be seen as a crack tip
that advances or recedes when the applied load is decreased or increased. However,
Sneddon's solution implies no interfacial stresses while as shown by Goodman[9],
when two bodies are pressed one against other, their surfaces undergo radially
inward tangential displacement, and with dissimilar materials the displacement
will be different for the two surfaces and this slip will be resisted by interfacial friction. The slip behaviour between two bodies is governed by Spence's
parameter (ref.10) which depends on the Poisson's ratios of the elastic solids.
Fortunately this parameter cancels out and the results are rigorously exact, for
the contact between a rigid body and a solid with ν = 0.5, i.e., a rubber-like
material.

Determination of the strain energy release rate G and of its derivative
($\partial G/\partial A$) with respect to the contact area makes it possible to determine the equilibrium and stability conditions of the system and to predict the kinetics of its
evolution under fixed or cyclic load, fixed grips, or fixed cross-head velocity
conditions (refs.11-13).

2 ADHESIVE CONTACT

The edge of the contact area, like any three-dimensional crack, is subjected
locally to a plane strain state (ref. 14), so that the strain energy release

rate is given by :

$$G = \frac{1-\nu^2}{2\ E} \cdot K_I^2 \qquad \qquad \ldots 1.$$

where K_I is the stress intensity. The factor 1/2 is introduced to allow for the rigidity of the indenter applied against the elastic solid. At equilibrium, $G = w$ (Griffith's criterion), where w is the Dupré work adhesion, determined from the surface and interface energies of solids 1 and 2 in contact, by : $w = \gamma_1 + \gamma_2 - \gamma_{12}$.

Let P_1 be the apparent load that, if the Sneddon integration constant $\chi(1)$ is zero, gives the same contact radius a as under the load P with $\chi(1) \neq 0$. We then obtain $K_I = (P_1 - P)/\sqrt{4\pi a^3}$, and for a spherical punch (ref. 8) :

$$G = \frac{(P_1 - P)^2}{6\pi a^3 K} \qquad \qquad \ldots 2.$$

P_1 being linked to the radius of contact a by :

$$P_1 = a^3 K/R \qquad \qquad \ldots 3.$$

From Sneddon's equations, we can derive the elasticdisplacement δ, also given in ref. 4 :

$$\delta = \frac{a^2}{3R} + \frac{2P}{3aK} \qquad \qquad \ldots 4.$$

hence $G = \dfrac{3K}{8\pi R^2 a} (a^2 - R\delta)^2 \qquad \qquad \ldots 5.$

At equilibrium under the load P_o, $G = w$, so that eq. 2 is the equilibrium relation derived by Johnson et al.[4], on the basis of the energy balance, and the equilibrium radius a_o is given by :

$$a_o^3 = \frac{R}{K} \left[P_o + 3\pi wR + \sqrt{6\pi wRP_o + (3\pi wR)^2} \right] \qquad \qquad \ldots 6.$$

The term in brackets is the correction to Hertz's theory that must be taken into account when the applied load P_o tends toward zero. For example, at zero load, there is a finite contact area with a radius such as $a_{p=0}^3 = 6\pi wR^2/K$.

Vertical displacement u_z of the points of the surface of elastic solid outside the contact area at the distance r from the center, may be deduced from Sneddon's equation (ref.6). It is shown in ref.8 that at the edge of the area of contact du_z/dr is infinite, so that the junction of the solid to the spherical punch is vertical : here we have the geometry of fracture mechanics as displayed on Fig.1, and this surface profile remain valid out of equilibrium ; the hertzian profile is given for comparison (light line).

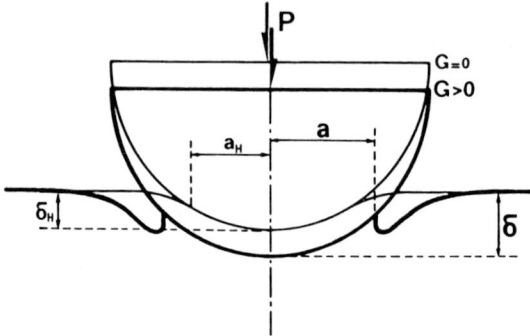

Fig. 1   Geometry of an adhesive contact in equilibrium condition for a rigid sphere on an elastic half-space. For comparison, Hertzian contact under the same load is also given (light line).

The stresses inside the elastic solid can be obtained (ref.8) by superimposing the hertzian compressive stresses $(\sigma)_H$ and a main pressure $(p_m)_H = P_1/\pi a^2$; and the tensile stresses of a flat punch $(\sigma)_F$ and a mean pressure $(p_m)_F = (P_1-P)/\pi a^2$. Normalizing with respect to mean pressure $p_m = P/\pi a^2$, the adhesive stresses may be written :

$$\frac{\sigma}{p_m} = \frac{P_1}{P}\left(\frac{\sigma}{p_m}\right)_H - \left[\frac{P_1}{P} - 1\right]\cdot\left(\frac{\sigma}{p_m}\right)_F \qquad \ldots 7.$$

Of course, this equation gives back exactly the Hertzian stresses if $P = P_1$.

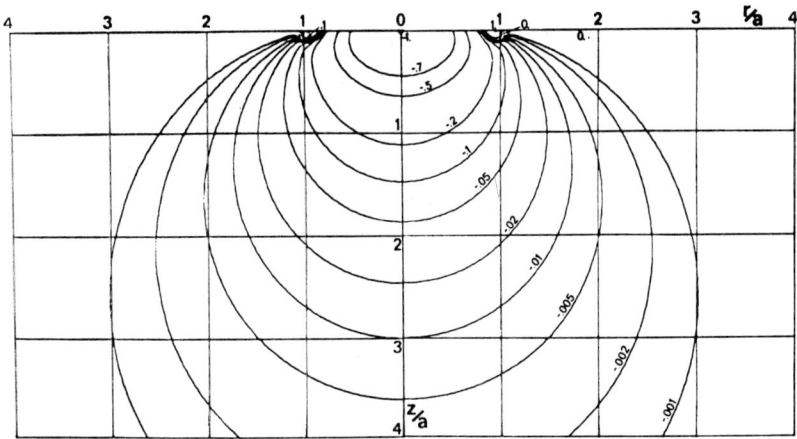

Fig. 2   Adhesive contact of spheres: principal stress values for σ_3, for $P=0$ and $\nu=.5$, normalized by $P_1/\pi a^2$.

For example, fig.2 shows, in reduced coordinates (ref.3), the principal isostress σ_3 for the special case $P = 0$ and $\nu = 0.5$, with the particular normalization $p_m = P_1/\pi a^2$. The existence of molecular attraction forces produces stresses in the bulk that become infinite at the edge of contact.

Let us consider (fig.3) the equilibrium of a spherical punch in contact with an elastic half-space under the load P_o. Radius of contact a_o and corresponding displacement δ_o are given by eqs.6 and 4 respectively. The stability of this equilibrium can be studied with a testing machine of finite stiffness k_m (ref.15), by turning the screw inorder to impose on the system a fixed displacement Δ. This is divided into the elastic displacement δ_m of the spring and the elastic displacement δ of the solid in contact with the rigid punch. Thus, a force $P = k_m \delta_m$ is exerted by the spring on the elastic solid and taking into account initial loading, one can write :

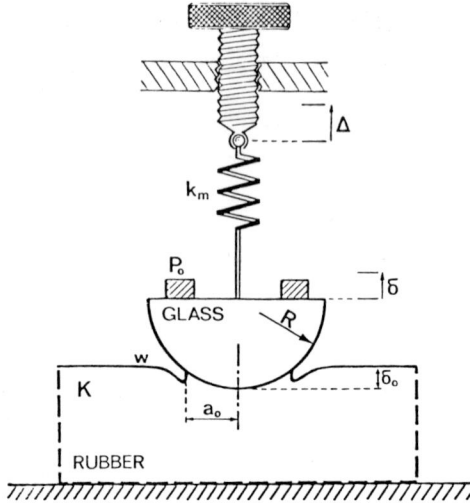

Fig. 3   Principle of the apparatus to study the state of equilibrium of a
rigid sphere in contact with an elastic solid, with a testing
machine with a finite stiffness.

$$\Delta = \frac{P-P_0}{k_m} + \delta-\delta_0 \qquad \dots 8.$$

and using eq.4, eq.8 becomes :

$$\delta = \left[\frac{a^3K}{2Rk_m} + \Delta + \frac{P_0}{k_m} + \delta_0\right] \Bigg/ \left(1 + \frac{3aK}{2k_m}\right) \qquad \dots 9.$$

Equilibrium at fixed Δ is given by $G = w$, and the limit of stability by $(\partial G/\partial a)_\Delta = 0$
which depends on the stiffness k_m. Courtel et al[16] have been shown that the sta-
bility range increases monotonically with the stiffness, from the fixed load case
($k_m = 0$) to the fixed grips case ($k_m \infty$). Introducing the reduced values (ref.4) :

$$\| P \| = P/3\pi wR$$

$$\| a \| = a / \left[\frac{3\pi wR^2}{K} \right]^{1/3}$$

$$\| \delta \| = \delta / \left[\frac{\pi^2 w^2 R}{3K^2} \right]^{1/3}$$

and adding : $\| \Delta \| = \Delta / \left[\frac{\pi^2 w^2 R}{3K^2} \right]^{1/3}$

and : $\| k_m \| = k_m / \left[\frac{3a_0 K}{2} \right]$

the stiffness of the spring being normalized by the stiffness of the initial con-
tact, one can represent the state of the system on a $\| \delta \| - \| a \|$ diagram (fig.4).
The locus of equilibrium points is the curve $\delta(a)$. For comparison, the case of a
non-adhesive contact is plotted (curve δ_H, Hertz's theory). Curves $(\delta)_\Delta$ show the
variation of $\| \delta \|$ with $\| a \|$ at fixed displacement $\| \Delta \|$; these are independent
of the Dupré work of adhesion and their minima, when they exist, are all located
on the Hertz curve δ_H. The limit of stability is represented by the point B
(ref.15). For a test at fixed load ($k_m = 0$) point B tends toward point C such that
$\| a \|^3 = 1/2$ and at fixed grips condition ($k_m = \infty$) it tends toward point D with
$\| a \|^3 = 1/18$ (ref.4, 11).

3 KINETICS OF ADHERENCE

The equilibrium defined by G = w may be disrupted by a change in displacement Δ, or in load if $k_m = 0$. When G > w, the two solids, elastic body and rigid
punch, begin to separate and the contact area decreases ; conversely, if G < w,
the contact area increases and the crack recedes. The difference G - w represents
the force applied per unit lenght of crack ; this crack takes a limiting speed v
that depends on the temperature. If it is assumed that viscoelastic losses are
proportional to w (refs. 17, 18), and localized at the crack tip, we may write
(ref. 11) :

$$G - w = w\phi(a_T v) \qquad\qquad ...10.$$

Where the second term is the drag due to viscoelastic losses at the crack tip,
a_T being the shift factor in the William-Landel-Ferry transformation. ϕ is a di-
mensionless function of the speed of propagation and the temperature. This function ϕ is characteristic of the viscoelastic material for the propagation in mo-

de I, and quite probably directly linked to the frequency dependence of the imaginary component of the Young's modulus. It has been shown (refs. 11 - 13) that ϕ is independent of the loading system and of the geometry. Knowledge of the function ϕ makes it possible to predict the evolution of the contact in all circumstances. The prediction assumes only that the rupture is adhesive, i.e, that the crack propagates at the interface, and that viscoelastic losses are limited to the crack tip, meaning that gross displacements must be purely elastic. As already pointed out, the interest of the proposed formula (eq. 10) is that surface properties (w) and viscoelastic losses (ϕ) are clearly decoupled from the elastic properties geometry and loading conditions that appear in G.

3.1 Crack propagation at fixed displacement Δ

Starting from equilibrium under the load P_o (point L in fig.4), let us apply, at the time t = 0, an instantaneous displacement Δ ; the equilibrium is then disrupted, the strain energy release rate (eq. 5), with δ given by eq. 9, increases and the contact area decreases as the crack advances. It is shown in fig.4 that there is first, from the equilibrium point L, an instantaneous modification of displacement δ at constant contact radius a_o (branch LM or LM'), corresponding to the elastic response of the elastomer. This is followed by a simultaneous variation of δ with a at fixed displacement Δ along the curve $(\delta)_\Delta$, that can lead to a new equilibrium state (branch MN) if Δ is greater than the critical value Δ_c, depending on k_m (ref.15), or to rupture (branch M'Q'), if not. The crack propagation speed increases or decreases according to whether the sign of $(\partial G/\partial a)_\Delta$ is negative or positive. The locus of points corresponding to the minimum of crack speed is plotted on fig.4 (curve OBJ).

The experiments were carried out using apparatus, shown in fig.5, consisting essentially of a precision balance supporting at the end of the arm, a hemispherical glass lens of radius R = 2.19 mm. This indenter is applied for a duration t_A = 10 min, under a compressive load P_o, against the flat surface of a polyurethane plate (VISHAY PSM4, K = 8.9 MPa, ν = 0.5). An instantaneous elongation Δ is then imposed by using a helical spring of stiffness k_m, attached at the other end of the balance arm and linked to a vertical micrometric stage. Five springs with stiffnesses 25.6, 41.9, 57.1, 84.3 and 122.6 N/m have been used. The contact area, illuminated by reflection of monochromatic light, is observed through the lens with a microscope. For a quantitative evaluation of the evolution of the contact area during rupture, a 16 mm camera records the contact areas at 25 frames per second with approximately tenfold magnification ; the frames we-

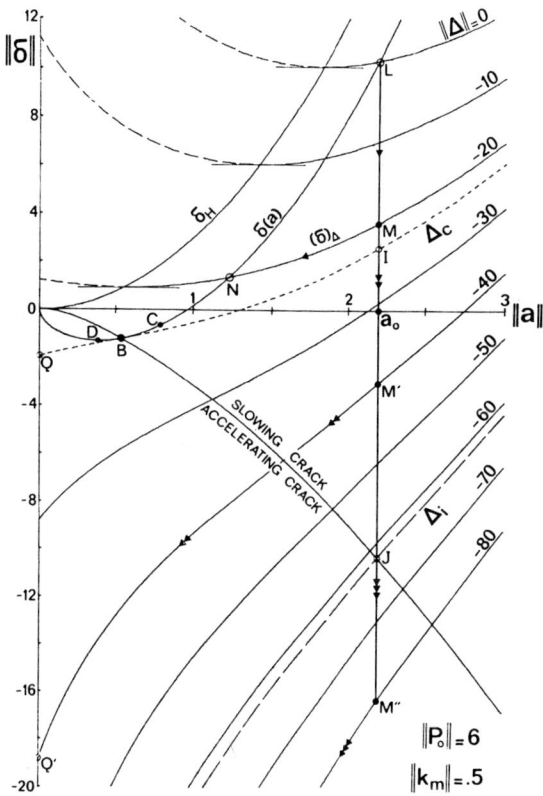

Fig. 4   Relations between elastic penetration δ and radius of contact
           area a of a rigid sphere in contact with an elastic body, in
           reduced coordinates. The equilibrium curve is δ(a); curves (δ)$_\Delta$
           show the variation of δ with a at constant displacement Δ. The
           curve δ$_H$ is given by the Hertz theory.

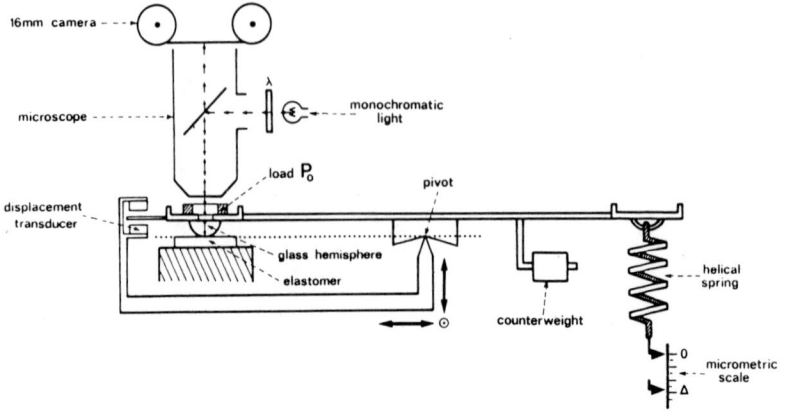

Fig. 5   Schematic arrangement of the apparatus used to study the adherence
of a glass hemisphere to viscoelastic materials.

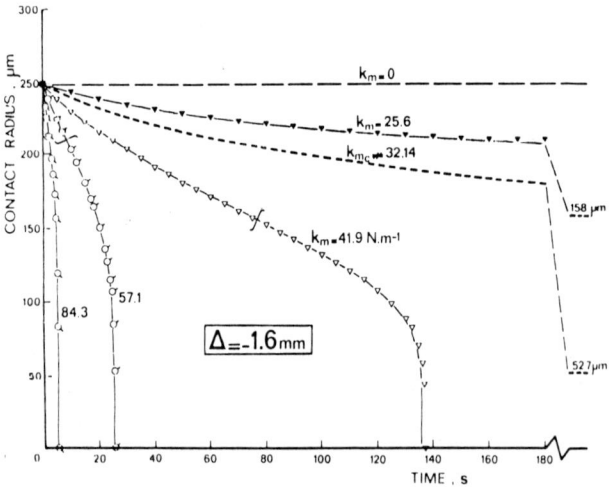

Fig. 6   Variation with time of the radius of contact area of a glass
hemisphere in contact with a polyurethane surface, for the same
displacement Δ and various stiffnesses k_m.
(P_o=50mN, R=2.19mm)

re then enlarged and measured. The apparatus also includes an inductive displacement δ of the indenter in the elastomer. Temperature and humidity relative remained constant for each set of experiments.

Figure 6 shows the variation of the contact radius with time for various stiffnesses, and a constant displacement $\Delta = -1.6$ mm. For small stiffnesses, a slow return towards a new equilibrium is observed with a continuously decreasing crack speed.As expected , for higher stiffnesses, a slowing down of the crack followed by an acceleration towards rupture is obtained, and the experimental points of inflexion are in good agrement with theoretical predictions (ref. 15). For comparison, the dashed line ($k_m=0$) represents the case of an unloading at constant load, and the a-axis, an unloading at fixed grips condition ($k_m=\infty$). Thus, fig.6 clearly shows the infuence of machine stiffness on the kinetics of adherence of an elastomer.

Equilibrium rupture condition for various stiffnesses k_m, initial loads P_0 and fixed displacement Δ are shown in a G-v diagram on fig.7. The strain energy release rate G is calculated by eq.5 with δ given by eq.9, and the crack speed v is derived from the slope at each points on the a(t) curves as in fig.6. All the results corroborate the previous findings concerning adherence of spherical or flat punches and peeling in a test at fixed load (ref.11), and consequently, confirm the general equation (eq.10) in which the function ϕ, for a polyurethane specimen, varies as the 0.6 th power of the crack speed. Moreover the graphs of displacement δ show that this quantity is elastic, and that viscoelastic losses are limited to the crack tip.

3.2 Tackiness of elastomers

When the dissipation function ϕ is known, it is to be expected that eq.10 can be used to predict the kinetics of evolution of the system if a fixed crosshead velocity is applied to the spherical indenter, with aid of a tensile machine This test of contact rupture is routinely used in the rubber industry to evaluate the tackiness of elastomers, a term that characterizes their peculiarity of adhering instantaneously to solid surfaces with which they are placed in contact. The maximum value of the force recorded during separation serves to characterize the tackiness of the material under the experimental conditions chosen. In fact, the force of adherence so determined depends not only on the intrinsic properties (surface and viscoelastic) of the solids in contact, but also on such experimental parameters as the velocity of separation, the initial applied load, the duration of contact under this load, the temperature and the relative humidity.

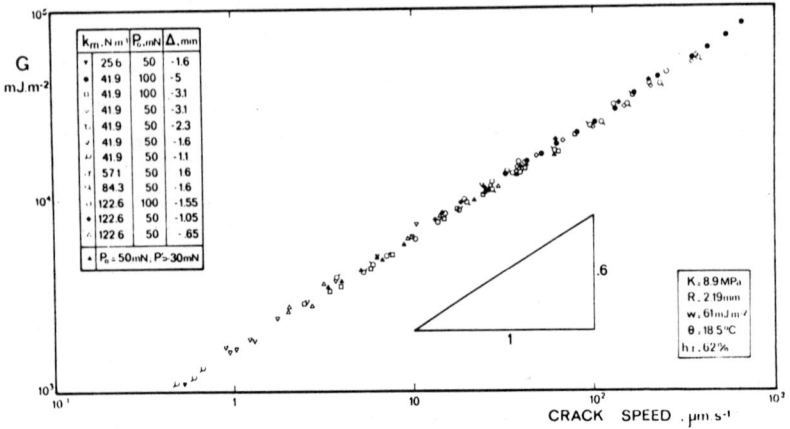

Fig. 7   Relation between the strain energy release rate G and the crack
          speed v=da/dt, for different tests with various stiffnesses,
          initial loads and fixed displacements Δ.

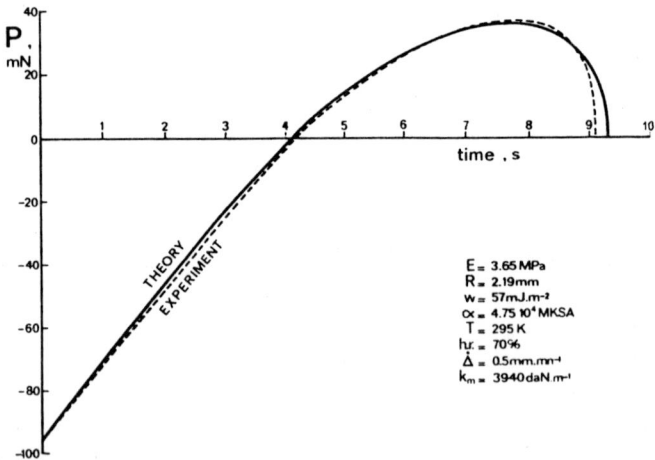

Fig. 8   Adherence force versus time for a glass ball in contact with a
          polyurethane surface, at fixed crosshead velocity. Comparison
          between computed and experimental curves.

Taking into account the stiffness k_m of the machine, if it cannot be regarded as infinite in comparison with the stiffness of the initial contact, it is shown in ref.19 that the velocity $\dot{\delta}$ of the indenter is related to the velocity $\dot{\Delta}$ imposed on the machine by :

$$\dot{\delta} = \left[\dot{\Delta} - 3K \ (\delta-a^2/R)v/2k_m\right] /(1 + 3aK/2k_m) \qquad \ldots 11.$$

Using eqs. 4,5 and 10, and with the help of a computer, it is possible to calculate the variation with time of the force to which the two solids are subjected. Figure 8 compares the computed curve and the experimental results. At the beginning of separation, the increase in the force results from the increase of δ, whereas at its end the rapid decrease in the force is due to the accelerated reduction of the contact area. The good agreement between theoretical predictions and experiments makes it possible to anticipate the influence of various parameters, such as temperature, crosshead velocity and surface properties (ref.19). Thus it has been confirmed that the force of adhesion(tack force), like the energy dissipation (tack energy), increases with the increase in the cross head velocity, the decrease in the temperature wich increases the viscoelastic losses at the crack tip, and the increase in the thermodynamic work of adhesion that allows the interface to sustain higher stresses.

3.3 Dwell time effect

The increase in the adherence of elastomers with the contact time is often attributed to the macroscopic creep of the viscoelastic material, which both undergoes deformation and "wets" the surface of the indenter (ref.20) and/or the microscopic diffusion of the free ends of the polymeric chains across the interface (ref.21). The last process can be recognized by the linear variation of the adherence force with the square root of the dwell time under the initial load.

The effect of the dwell time on the adherence has been studied by carrying out several instantaneous unloading of a glass ball in contact with a polyurethane surface, from the same initial compressive load to the same final tensile load, for different contact times varying from 1 min to 15 hours (ref.22). We have recorded, as expected, the increase in radius of contact with dwell time, but we find that it tends towards a particular value, which is observed after a critical time equal to 10 min, whatever the initial load. One surprising result is the increase in the time required for rupture with dwell time beyond 10 min ; it varies as the 0.2 th power of the duration of contact, a phenomenon apparently incompatible with the equilibrium state thought to have been reached after 10 min, because of the absence of subsequent variation of the contact area. This fact has been interpreted on the basis of a decrease of the force applied to the crack defined by eq.10, resulting on first analysis in an increase in thermodynamic work of adhesion w with dwell time.

Figure 9 gives, for various times of application of the initial load, the relation between the strain energy release rate (eq.2 with eq.3) and the crack speed. The six curves obtained are, in log-log coordinates, straight and parallel, with 0.6 as the common slope. These results corroborate the previous findings as in fig.7, and their distribution is similar to that observed in peeling experiments in various liquids(ref.17) or on various substrates (ref.18), and in cylinder rolling experiments in atmospheres of varying humidity (refs13, 23). The vertical offset results from the variation in Dupré work of adhesion w with the nature of the liquid or the substrate, or with the water vapour content in air, thereby confirming the multiplicative effect of w on viscoelastic losses at the crack tip (eq.10). In assigning to every dwell time a particular value of w, one obtains the master curve characteristic of the polyurethane (refs 11-14).

The comparison between values of dwell time and of the associated work of adhesion shows that w varies as the 0.1 th power of the contact duration. Thus, since the rupture time t_R varies as the 0.2 th power to the contact duration, measurement of rupture time is a simple way of determining the work of adhesion from the relation $w \sim \sqrt{t_R}$.

The proportionality of w with the 0.1 th power of the contact duration, rather than with the square root, proves that the involvement of diffusion of the free ends of the polymeric chains across the interface cannot be maintained for glass-polyurethane contacts.Consequently, the increase in adherence with time may be attributed to a mechanism having a slower kinetics than that of diffusion, such as the relaxation of stresses in the roughnesses of the contacting surfaces. This is the reason why the decrease in the force applied to the crack with time must be ascribed to a decrease in the strain energy release rate G, w remaining constant, due to the relaxation of the additional elastic energy stored in the asperities that are highly compressed during the initial loading stage, this energy being not taken into account in the computation of G. In ref. 22 a simplified calculation of this energy confirmes this assumption .

Although incorrect, the hypothesis of an increase in the work of adhesion with dwell time nevertheless has the advantage of predicting the evolution of the system when a rigid indenter is submitted to a cyclic loading, as in a fatigue test : a compressive load P is applied during a time t_A, then a tensile load P' is applied during a time t_R, and so on ,so as to produce a crack opening and closing cycle. The crack propagates faster in a contact zone where the compres-

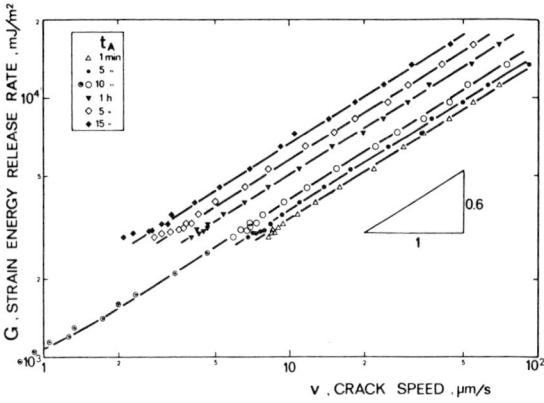

Fig. 9   Relationship between the strain energy release rate G and the crack
          propagation speed v for various dwell times.(P=50mN, P'=-30mN,
          R=2.19mm).G varies as the 0.6th power of v, as in Fig. 7.

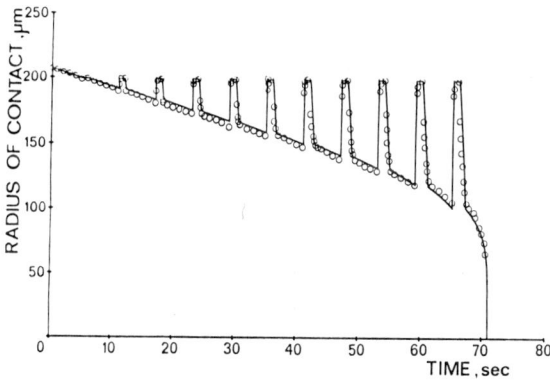

Fig. 10   Cyclic loading : Contact radius versus time for a glass ball in
          contact with a polyurethane surface.(P=30mN, P'=-30mN, R=2.19mm,
          t_A=1s, t_R=5s, w_1=57mJ/m^2, w_2=4.8mJ/m^2).

sive load has been applied for a shorter time (push-on phase), because of the existence of unrelaxed stresses in the roughnesses crushed in the contact area ; so, it is possible to observe the complete failure after a small number of cycles as shown in fig.10. Assuming that the particular mode of loading creates an experimental condition of kinetics between two apparent works of adhesion, w_1 corresponding to equilibrium and $w_2 < w_1$ for contact of short duration, one can predict (ref.3) the results of fig.10 by numerical integration of eq.10, taking into account eqs.2 and 3 (solid line). Experiments have shown that 1) the complete rupture of the contact area involving pull-off/push-on stages cannot be observed, if it does not occur when the tensile load is continuously applied ; 2) cyclic loading delays the failure ; 3) the observation of complete rupture of the contact depends on the time of application of the tensile force , i.e. for the shortest duration a stable cycle is obtained. Thus we can deduce that the fatigue of an adhesive joint cannot be observed if the material is purely viscoelastic, i.e. if no damage takes place around the crack tip.

4 ADHERENCE OF A RIGID FLAT ENDED SPHERE

Sneddon's solution (ref.6) for axisymmetric punches of arbitrary profile, already used for the adhesive contact of a rigid sphere, also allows us to study the adherence of a sphere of radius R, with a flat of radius a_F in contact with an elastic halfspace over an area of radius $a > a_F$. Letting $\alpha = a_F/a$, one can write (ref.24) :

$$P = \frac{3\ aK}{2} \left[\delta - \frac{a^2}{3R}\ (1-\alpha^2)^{3/2} \right] \qquad \qquad ...12.$$

$$P_1 = \frac{a^3 K}{2R}\ (2 + \alpha^2)(1-\alpha^2)^{1/2} \qquad \qquad ...13.$$

For $\alpha = 0$ or $\alpha = 1$, these equations give the relations between loads, penetration and radius of contact for an adhesive sphere or for a flat punch.

The strainenergy release rate is given by eq. 2, and the equilibrium is defined by $G = w$. Figure 11 gives the equilibrium radii of contact of flat-ended glass balls applied to a polyurethane surface under various loads for four value of a_F : agreement with theoretical curves (solid lines) is within 2 percent.

The kinetics of crack propagation were studied according to eq.10 ; for example, fig.12 compares the variation of radius of contact area with time, for the same unloading applied to a glass ball and to a flat-ended glass ball. Solid lines are theoretical curves obtained by numerical integration of eq. 10. Agreement is quite good, and the change in regime is clearly seen for $a < a_F$. It is interesting to observe that a small flat on a sphere can markedly delay the rupture of contact.

49

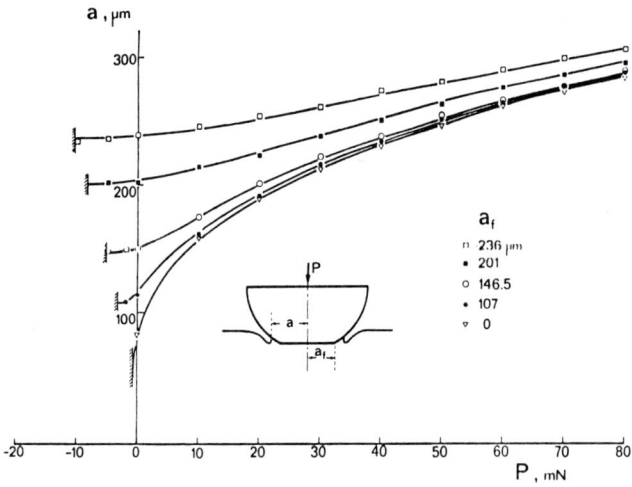

Fig. 11   Equilibrium radii of contact for four flat-ended spheres in contact
           with a polyurethane surface. Drawn lines are for theory.(R=2.19mm,
           K=8.9MPa, w=50mJ/m^2).

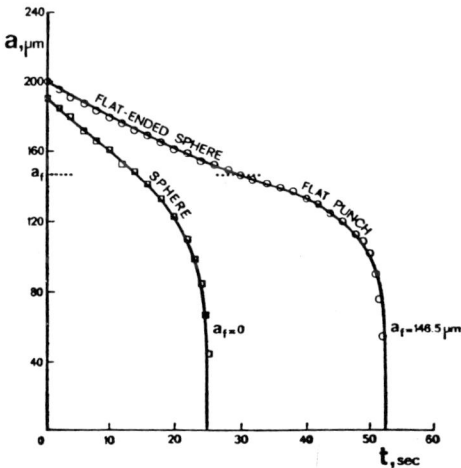

Fig. 12   Flat-ended sphere:kinetics of unloading from P=20mN to P'=-30mN,
           for a_f=0 et a_f=146.5μm. Comparison with theory.

50

5 SLIDING FRICTION

When a rigid sphere slides over a surface of natural rubber or polyurethane, a surprising mode of deformation appears. This has been described by Schallamach[25] as "waves of detachment". These are small regular folds, filled with air,
which cross the contact area at high speed, from front to rear, i.e from a compressive zone to a tensile zone, and they move like wrinkles in a carpet (Fig.13) .The
formation of these "Schallamach waves" can be explained by the existence of a
particular profile of the surface at the edge of contact (geometry of fracture
mechanics) and a strong adherence that provoke a buckling in the front part of
the slider (ref.26). When a fold moves, the elastomer first peels from the rigid
surface and then sticks again. Assuming there is no slip between two consecutive
wrinkles, the friction force T is directly related to the peeling force F in the
front part of every fold (ref.27). If n folds are present in the contact area
of width 1, and propagate with a mean speed v, when the slider moves with a velocity V on the surface of elastomer, one can write the energy balance as (Fig.14):

$$V.T = n.F.v$$

where $F = G.1$, as in a $\pi/2$ - peeling test, G being the strain energy release rate. For the polyurethane surfaces studied, G varying as the 0.6 th power of the
crack speed, a variation of V.T/n.1 with the 1.6 th power of the wave speed shou
be observed. Figure 15, obtained for various velocities of slider and normal loads, agrees well with the theoretical prediction (ref.28). Consequently, if wave
of detachment appear in a sliding contact area, the friction force is effectively due to the total summation of the peeling forces in the front part of these
folds.

6 CONCLUSION

It is shown that the Sneddon solution of the Boussinesq problem allows one
to study the influence of molecular attraction forces on the contact of elastic
solids. The introduction of the concepts of fracture mechanics, such as the
strain energy release rate, enables one to define the stability conditions and
to predict the kinetics of adherence of elastomers. It also allows solution of
the problem of the tackiness of elastomers and enables one to take into account
the dwell time effect and the influence of the stiffness of testing machine on
the adherence.

ACKNOWLEDGEMENTS

The author would like to thank the Direction des Recherches, Etudes et
Techniques, for the financial support given to this work (DRET contrat n° 78-609

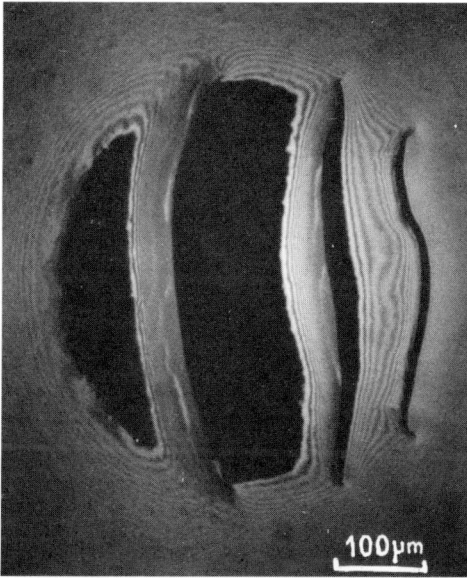

Fig. 13   Area of contact of a glass ball sliding on a polyurethane surface,
           illuminated with monochromatic light. Crack opening and closing
           modes are clearly visible.

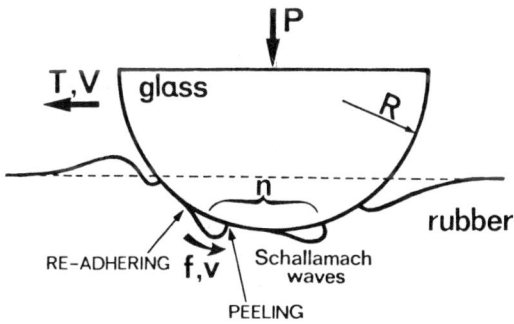

Fig. 14   Schematic diagram showing the profile of the polyurethane surface
           near the edge of contact, and the shape of Schallamach waves.

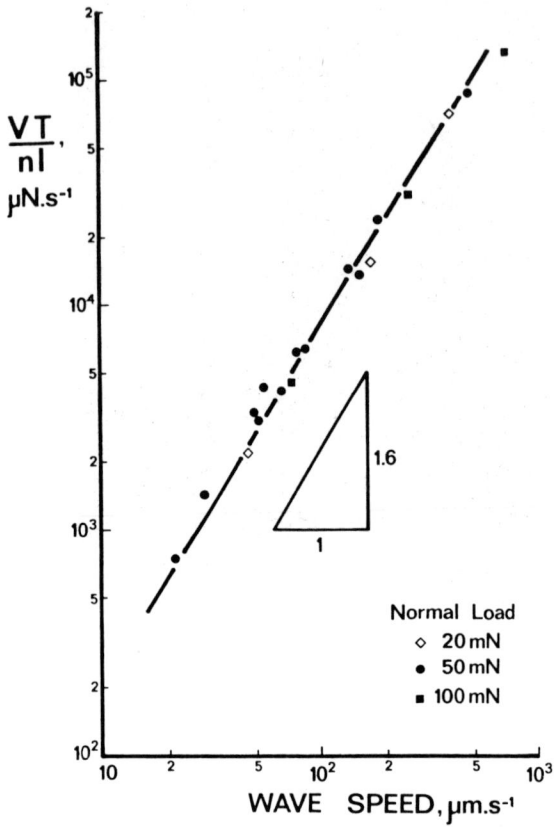

Fig. 15   Relationship between the friction force and the propagation speed of Schallamach waves.

REFERENCES

1. Hertz, H. "Über die verüh
rung fester elastischer Körper" J. für die reine und angewandte mathematik, vol. 92, 156-171, 1881.

2. Lawn, B.R "Hertzian fracture in single crystals with the diamond structure", J. Appl. Phys., vol. 39, 4828-4836, 1968.

3. Barquins, M. "Adhesive contact and kinetics of adherence between a rigid sphere and an elastomeric solid", Int. J. Adhesion and Adhesives, April 1983, in press.

4. Johnson, K.L., Kendall, K. and Roberts, A.D. "Surface energy and the contact of elastic solids", Proc. Roy. Soc. Lond., vol. A 324, 301-313, 1971.

5. Johnson, K.L. "A note on the adhesion of elastic solids", Brit. J. Appl. Phys., vol. 9, 199-200, 1958.

6. Sneddon, I.N. "The relation between load and penetration in the axisymmetric Boussinesq problem for a punch of arbitrary profile", Int. J. Engng. Sci., vol. 3, 47-57, 1965.

7. Maugis, D. and Barquins, M. "Adhesive contact of a conical punch on an elastic half-space" J. Phys· Lettres, vol. 42, L95-L97, 1981.

8. Barquins, M. and Maugis, D. "Adhesive contact of axisymmetric punches on an elastic half-space : the modified Hertz-Huber's stress tensor for contacting spheres " J.Méc· Théo. Appl., vol. 1, 331-357, 1982.

9. Goodman, L.E. "Contact stress analysis of normally loaded rough spheres", J. Appl. Mech (ASME), vol. 29, 515-522, 1962.

10. Spence, D.A. "The Hertz contact problem with finite friction", J. Elasticity, vol. 5, 297-319, 1975.

11. Maugis, D. and Barquins, M. "Fracture mechanics and the adherence of viscoelastic bodies", J. Phys. D : Appl. Phys., vol. 11, 1989-2023, 1978.

12. Maugis, D. and Barquins, M. "Fracture mechanics and adherence of viscoelastic solids", in Adhesion and adsorption of polymers, ed. L.H. Lee, vol. 12A, Plenum Publ. Corp., N-Y, 203-277, 1980.

13. Barquins, M. "Etude théorique et experimentale de la cinétique de l'adhérence des élastomères", Thesis, Paris,,1980.

14. Kassir, M.K. and Sih, G.C. "Three-dimensional stress distribution around an elliptical crack under arbitrary loadings", J. Appl. Mech., vol. 33, 601-611, 1966.

15. Barquins, M. "Influence of the stiffness of testing machine on the adherence of elastomers", J. Appl. Polymer Sci., 1983, in press.

16. Courtel, R., Maugis, D. and Barquins, M. "Note sur une formulation commune de certains problèmes d'équilibre de deux corps élastiques adhésifs", Industrie minérale, vol. 4, n° spécial Rhéologie, 137-143, 1977.

17. Gent, A.N. and Schultz J. "Effect of wetting liquids on the strength of adhesion of viscoelastic materials", J. Adhesion, vol.3, 281-294, 1972.

18. Andrews, E.H. and Kinloch, A.J. "Mechanics of adhesive failure", Proc. Roy. Soc. Lond., vol. A 332, 385-399, 1973.

19. Barquins, M. and Maugis, D. "Tackiness of elastomers", J. Adhesion, vol. 13, 53-65, 1981.

20. Bates, R. "Studies in the nature of adhesive tack", J. Appl. Polym. Sci., vol. 20, 2941-2954, 1976.

21. Koszterszitz, G. "Eigenklebrigkeit und tack von unvernetzten kautschuken", Colloïd and Polymer Sci. vol. 258, 685-701, 1980

22. Barquins, M. "Influence of dwell time on the adherence of elastomers", J. Adhesion, vol. 14, 63-82, 1982.

23. Roberts, A.D. "Looking at rubber adhesion", Rubber Chem. Techn., vol. 32, 23-42, 1979.

24. Maugis, D. and Barquins, M. "Adhesive contact of sectionally smooth-ended punches on elastic half-spaces : theory and experiment", J. Phys. D : Appl. Phys., to be published.

25. Schallamach, A. "How does rubber slide ?", Wear, vol.17, 301-312, 1971.

26. Barquins, M. and Courtel, R. "Rubber friction and the rheology of viscoelastic contact", Wear, vol. 32, 133-150, 1975.

27. Briggs, G.A.D. and Briscoe, B.J. "How rubber grips and slips : Schallamach waves and the friction of elastomers", Phil. Mag., vol 38, 387-399, 1978.

28. Barquins, M. "Energy dissipation in Schallamach waves" Wear, 1983, in press.

Chapter 4

STATIC AND DYNAMIC CHARACTERISTICS OF BONDED JOINTS AS AFFECTED
BY AGEING AND ENVIRONMENTAL CONDITIONS

A.A. Khalil and M.M. Sadek

Department of Mechanical Engineering, University of Birmingham, U.K.

ABSTRACT

The feasibility study carried out on the use of Epoxy resin bonding in
fabrication of machine structures has been proved successful as applied to
the manufacture of both a prototype horizontal milling machine and a large scale
sealed cover of a motor unit. This study has shown that the bonded milling
machine is considerably better than the commercial one both in its dynamic
performance and the rate of metal removal.

At this stage, it is essential to furnish the designer with the data
required for such type of fabrication. The aim of this project is to determine,
experimentally, the static and dynamic characteristics of bonded joints of the
type used in the above mentioned study under various types of loading. Also the
effect of curing time, ageing and environmental conditions on the strength of
these joints have been examined. Comparison between four types of adhesives
have also been carred out.

As a result of this investigation, it has been found that a 6% reduction
in the joint strength resulted from ageing of twelve weeks. If however, the
joint has been immersed in hydraulic oil, no strength reduction was noticed.
The immersion of the joint in tap water, sud oil or salt water, caused the
strength reduction, in 12 weeks, to approach 30%. Maximum joint strength was
achieved after a curing period of eight days.

Fatigue tests were carried out on bonded specimens, showing an increase
in fatigue life of specimens with the curing period, until a maximum is reached

after almost a week. Beyond this point the bond fatigue life decreases at a
slow rate. The S-N curves show that the endurance limit is in the range
11×10^6 M/m^2.

1. INTRODUCTION

The traditional method of casting has several inherent disadvantages:
section thicknesses are governed by the prerequisites necessary for casting
complex structures rather than by the pure strength requirements. Production
flow is slow, relatively inflexible, and repair and painting of castings form
a significant part of the production cost.

Although the manufacture of machine structures by fabrication reduces
the above mentioned disadvantages, it has not been considered seriously in the
past since the available techniques have drawbacks in practical application.
For instance, welding which has many attractions, has the disadvantage of
causing structural distortions, a major drawback when accuracy is of prime
importance. In addition, welded structures may require heat treatment, an
unwelcome extra production operation, especially in the case of large machines.
Other fabrication techniques such as bolting and rivetting are also impractic-
able and cause unequal stress distribution in the joint. Adhesive bonding can
be used to good advantage in the manufacture of machine tool structures (1)
and by combining bonding with other manufacturing techniques, a new dimension
is added to the available fabrication methods.

When adopting the bonding technique the following possibilities arise:
Metal thickness need not be greater than is necessary for strength.
Arrangements which not only give stiff and light structures but also
give better accessibilities for maintenance.
Different materials may be used in the structure to improve their
properties (2).

This technique also allows adaptability for modularisation and therefore
the large-scale production of standard machines as well as the one-off production of specialised machines, Thus the bonding technique adds an important
contribution to the feasibility of interchangeable modules for machine structure

11. USE OF STANDARDISED JOINTS

Tha main features of the bonding technique is the use of standardised
joints and steel plates in making up the structure. Joints must be designed

for bonding. By following this principle much better joints are achieved than if bonding simply is adopted as a substitute for welding in a joint designed for welding. It is well known that bonded joints perform best under tension, compression or shear loading: less well under cleavage and relatively poorly under peeling loading.

Initially, the overarm of a milling machine was fabricated using British Standard channels and I-beams. The extension of this method to the fabrication of other structures has the advantage of easily obtainable standard parts. However, since the dimensions of web thickness of these parts are closely related, it meant compromise on the part of the designer on metal thicknesses, thus leading to overdesign, a major objection to the casting technique. This problem may be minimised by the use of non-standard angles, channels and I beas beams, thus removing the main advantage of these joints.

The use of standard angles, channels and I-beams was found to be impractical, thus double-containment joints shown in Figs. 1a and 1b were utilised. The modified angle and T-joints shown in Figs. 1c and 1d, made better use of the metal, though at the expense of joint symmetry. An optimum double-containment joint is one that is designed in such a way as to make the load-carrying capacity of the two webs and the adhesive layer equal to that of the plate to be bonded. To obtain this optimum strength the adhesive thickness should be uniform throughout the joint and equal to optimum values.

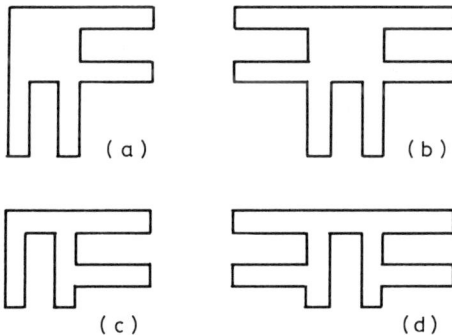

Fig. 1.
Double containment angle and T joints, symmetrical and compact styles.

Four commercial processes were investigated to find the most suitable method of manufacturing double-containment joints: extrusion, hot rolling and forming, continuous casting and cold forming. By far the most promising method of manufacture was extrusion, offering accuracy and flexibility as well as economy in the production of small quantities (3). To change joint design, only a simple die needs to be changed at a low cost.

Aluminium joints are ideal in many ways, however their use may, in certain circumstances, be limited as a result of possible corrosion or differential thermal expansion if bonded to steel or other materials.

III   BONDING PROCEDURE FOR LINE PRODUCTION

The bonding procedure can have a marked effect on the strnegth of bonded joints and a well-controlled operation is therefore essential. Before the application of the adhesive it is vital to ensure that the adherent surface is clean and free from foreign matter that may impair joint strength. Furthermore a chemical pretreatment of the adherent surface may be necessary in some cases. For instance pickling aluminium in a solution of chromic acid and sulphuric acid ensures that the bond strength does not deteriorate with time. Excessive rust has to be removed by other methods, such as shot blasting.

Since adhesives have to be used quickly after preparation, the method of manual mixing and application is obviously unsuitable for line production. However, dispensing machines are available for this type of application, where bonding using Robots, an added attraction to the bonding technique.

Although curing of the adhesive at room temperature is convenient, high-temperature curing is advantageous. Not only is a higher bond strength achieved but there is also a reduction in the curing time - very useful for line production.

IV   INDUSTRIAL APPLICATIONS

Four applications of bonding techniques have been carried out in collaboration with various industrial establishments. These are summerized as follows:

1.   Fabrication of A Bonded Overarm of a Horizontal Milling Machine

The box overarm of a small horizontal milling machine was chosen on two grounds. Firstly, both a cast iron and a fully welded component of the same type were already available, which made possible a comparison of the three

manufacturing methods, Secondly, the particular horizontal milling machine had
been tested on previous occasions (2) and its dynamic behaviour was fully investigated.

The original cast iron cross-section of the overarm, shown inFig. 2a,
was modified into the form shown in Fig. 2c. The figure also contains the
cross-section of the welded overarm, see Fig. 2b. The cross-sectional area A
and its second moment "Igg" for the three designs are also given in the figure.
Note that, for the bonded component, both were substantially larger than for the
welded design, mainly because of the central I section (Fig. 2c). This was
quite deliberate since, on the basis of the pilot study, it was considered to
be desirable to compensate for the loss of stiffness due to the lower shear
modulus of the bonding agent, a precaution proved to be unnecessary. The
dynamic characteristics of the redesigned overarm in relation to cast and
welded components were assessed, using forced vibrations as well as cutting tests.

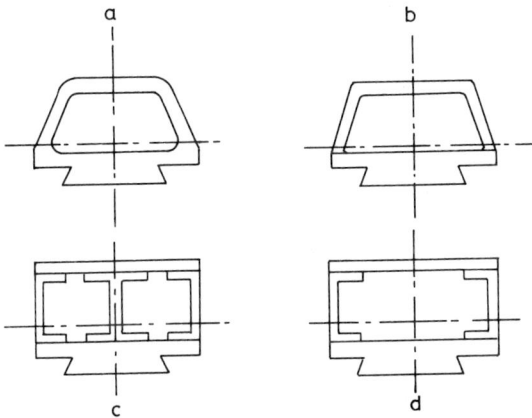

Fig. 2.
Cross-section of the different overarms:

A   cast,   B   welded,   C   bonded,   D   modified bonded

The relative improvement coefficient relating the performance of the milling machine with the bonded overarm to that with the case iron overarm is given by the rate of metal removal of the former case in relation to that of the welded or cast iron overarm as tabulated in Table 1.

TABLE 1   Improvement of Rate of Metal Removal of Bonded Overarm

	Bonded/welded	Bonded/cast iron
Improvement in rate of metal removal	18%	48.2%

It has already been mentioned that the cross-section and the weight of the bonded overarm were greater than those of the other two, (4). A summary of the weights of the various overarms are given in Table 11 showing that the modified bonded overarm is the lightest.

TABLE II   Weight of overarms

	Cast Iron	Bonded	Welded	Modified bonded
Total weight Kg.	69	83	37	42

It might well be concluded that the bonded overarm performed better than the other two due to its greater static stiffness. This was not the case. Quite to the contrary; its equivalent static stiffness, which counts from the vibration point of view, was lower than those of the other two. However, at the same time, its equivalent damping factor was very much higher and because of the interplay of these two, a better performance was obtained.

It was found that the natural frequencies of the three overarms were very close to each other, in spite of their differing static weight, as given in Table II. The equivalent static stiffness of the bonded overarm was 74 per cent of the cast and 81 per cent of the welded type. However, the damping factor of the bonded was 171 per cent of the cast and 175 per cent of the welded component.

Realistic costs are difficult to produce, since they depend on the scale of production, i.e. whether or not special jigs are justified. On a one-off basis the production costs of the four types of overarm are estimated in Table IV. These figures (produced in 1975) cover material, labour costs and overheads. For the cast iron components the cost of the pattern (£20) was excluded. When produced in batches the cost of the cast-iron overarm drops to £33. The cost of the welded type is not very much affected by numbers since this is a highly labour-intensive process, in which relatively little can be saved by special jigs. Roughly the same considerations also apply to the bonded type, although the preparation of surfaces for bonding and the application of the adhesive can be mechanised.

TABLE III   Estimated Production Cost of the Four Types of Overarm

Overarm	Cast iron	Welded	Bonded	Modified Bonded
Cost	£77	£68	£45	£40

These cost figures cover production only. They do not take into consideration stock levels required for avoiding delays in delivery, which are particularly high for cast components and low for bonded ones. Bottlenecks in the supply of highly qualified labout, as required for welding, may also constitute an important consideration. The stress-relieving heat treatment required for welded components may also be a bottle-neck because of limited furnace capacity, either of throughput or of size. This suggests that bonding will be particularly attractive when the components are very large and/or when a large number of them is required at short notice.

2. Design of a Prototype Horizontal Milling Machine

The feasibility of the technique having been established, it was applied to the manufacture of the base, column and overarm of a horizontal milling machine. The drives, knee, spindle, etc., were unchanged and hence the dimensions of the structural parts that interface with the unchanged parts were identical to those of a production model. Fig. 3 shows the main features of the prototype. As this was to be manufactured on a one-off basis, all the double-containment joints and the V-ways were machined from solid.

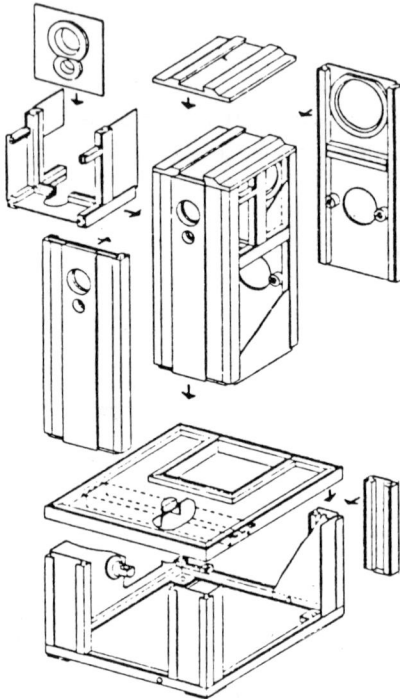

Fig. 3.

Isometric drawing showing the main features of the
prototype bonded milling machine.

In Fig. 4a a photograph of the bonded horizontal milling machine is shown
while the commercial milling machine is given in Fig. 4b for comparison of the
shape of the prototype in relation to the production type.

The dynamic performance of the bonded milling machine was evaluated by
determining the rate of metal removal for a wide range of depths of cut and
various cutting configurations. This illustrates that the improvement in the
rate of metal removal achieved for the bonded milling machine is within the

Fig. 4.

Horizontal milling machine: A – bonded, B – cast.

64

range of 60-500% in the case of up-milling as compared to the production machine. In the case of the down-milling the improvement in the rate of metal removal is within 200-700%.

To verify the above mentioned dynamic performance, cutting tests were carried out on mild steel specimens. A sample of the cut on both the bonded and the cast-iron machines is illustrated in Fig. 5, showing a considerable increase in the width of cut at which the machine starts to chatter (5).

3. Large Scale Drip Proof Cover for Prototype I Range Motor Unit

A drip proof Cover for a prototype motor unit has been fabricated by bonding and the production costs were compared with the welded one. The adhesive used was Araldite 2004. This investigation showed that a significant cost saving is possible using the adhesive bonding technique. For this cover it is estimated to be 11% of the total cost.

4. Fabrication of Spiral Carbide Tipped Endmills by Bonding

This was a collaborative work with Marwin Cutting Tools Company and the S.E.R.C. to develop a bonding technique for bonding the carbide tips to the end mill shank as a replacement of brazing process. This investigation (6) showed that the cutting performance of the bonded end mills is identical to that of the commercially available brazed ones in both cases of normal cutting conditions. Infrared prints demonstrated a pronounced heat isolation due to low thermal conditions and conductivity of the adhesive.

V. STRENGTH OF BONDED JOINTS

For rational design of structures fabricated by bonding, the designer should be furnished with the static and fatigue characteristics of both standard and industrial bonded joints. Besides, a simple mathematical formulae should be available to predict the dimensions and geometry of the bond line and the adherends. This paper will concentrate on the former requirement. Static and fatigue tests were carried out on single and double lap joints to assess the effects of the changes in either the joint geometry of the adherend's surface roughness on the joint strength under different types of loading.

i. Types of Joints

The tests were carried out on two types of specimens:

1. Standard Specimens: such as single and double lap joints, Butt and Modified joints, shown in Fig. 6 and tabulated in Table IV.

Fig. 5.

Machined specimens:
   a.   from bonded machine, maximum width of cut 42 mm.
   b.   from cast iron machine, maximum width of cut 18 mm.

Single-lap joint

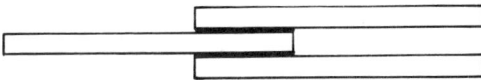

Double-lap joint

Butt joint

Fig. 6.
Standard joints used.

TABLE   IV

TYPE OF LOADING	TYPE OF JOINT	TESTING M/C	RANGE
STATIC PREDOMINANTLY SHEAR (some tensile)	Single lap	Instron tensile testing machine	l=10,15,20,25,30 mm t=0.81,0.91,1.22,1.63,1.96, 2.55 mm δ=50-500μm L=112 mm, b=25mm Sand blasted abraded-smooth
	Double lap	Instron tensile testing machine	l=10,15,20,25,30 mm t=1.63, 1.96 mm δ=50-500 μm L=112 mm; b=25 mm Surface condition-sand blasted smooth
PURE TENSILE	Butt	Denison tensile testing machine	
FATIGUE TESTS PREDOMINANTLY SHEAR (some tensile)	Single lap	Amsler frequency vibrophone	l = 10, 15, 20 mm
PURE TENSILE	Butt	Amsler frequency vibrophone	

2.   Industrial Joints: these are double containment T Joints shown in
Fig.7, which were used in the fabrication of the prototype horizontal milling
machine and the motor unit cover.   In this type of joint four adhesive   agents
were used to compare their effects on the joint strength.

ii.   Preparation of Bonded Joints

Surface Treatment: In all cases excessive grease has been wiped off with
a cloth dipped in chlororoethyene liquid prior to the vapour treatment.   All
specimens used with Belzona were shot blasted according to their specifications.
The other specimens were bonded with good machine finishes.

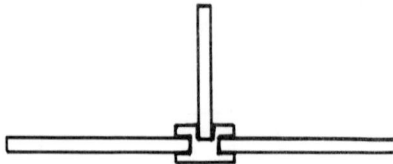

Fig. 7.   T - joint.

The mixing of the bonding agents and their curing times are highly dependent on the type of adhesive used. Those used were:

1. Araldite 2004

    Mixing ratio: 100:40
    Recommended curing time: 48 hours at room temperature

2. Bostic M890

    Acrylic type adhesive, it is an reactive bonding system consisting of an activator which is applied to only one of the two surfaces to be bonded and an adhesive base which is applied to the other.

3. Belzona Molecular-Metal ("Super metal")

    Mixing ratio: 3 parts of base to one part of the solidi-fier by volume.

    Recommended curing time: 25 minutes at room temperature.

4. Belzona Metal E:

    Mixing time: one part base to one part solidifier by volume.
    Recommended curing time: six minutes at room temperature.

VI TEST RESULTS AND DISCUSSIONS

1. Strength of Standard Specimens
(a) Static Loading

The static tests on lap joints illustrated in Fig. 8a show that the mean failure shear stress increased with increase in the adherend thickness until a thickness of 2 mm is reached, beyond which the effect becomes insignificant. For a single lap joint, the effect of the adhesive thickness on the joint strength is illustrated in Fig. 8b. This shows that the strength increases with the adhesive thickness until an optimum thickness of 0.15mm is achieved, beyond which a further increase will reduce the joint strength.

The effect of the length of the lap on the shear stress is plotted in Fig. 8c, for various adhesive thicknesses, showing a decrease in the mean shear strength with the increase of the lap length. It also shows that the adhesive thickness at low lap lengths results in a maximum variation in the joint shear strength within 10% with the adhesive thickness. On the other hand the effect of the adhesive thickness is insignificant at large lap lengths.

In Fig. 8d the variation of the shear strength with lap length are plotted for double and single lap joints, for both smooth and shot blasted surfaces. This chart shows that the double lap joints are stronger than the single lap by about 70% due to the decrease in the peeling stresses in the former case, arising from the misalignment of the applied load. It can also be noted that the effect of the surface finish on the joint strength is insignificant in the case of the double lap joints.

The effect of the surface roughness is highly effective in the case of the single lap joint as can be seen from the lower two curves in Fig. 8d. It is notices that the sand blasting increases the strength of the single lap joint by about 30%. This increase is due to the fact that the cavities created by the blasting will be filled with the adhesive and thus increase bond surface area resisting the applied load.

Standard static tests were carried out to assess the characteristics of various types of adhesives. Butt type specimens were used to determine the tensile strnegth while single lap joints were utilized to detect the average shear strength. On the other hand, bulk specimens were used on the Izod impact testing machine to determine the fracture energy. The results of these tests are summarized in Table V.

TABLE V.

Type of Adhesive	Tensile test N/mm^2	Shear test N/mm^2	Impact test J/mm^2
Araldite 2004	13.6	11.3	0.063
Bostik M890	17.7	8.6	0.055
Belzona "Super"	10.8	8.0	0.059
Belzona "E"	8.5	5.1	0.070

(b) Dynamic Loading

The fatigue tests were performed on the Amsler machine of maximum capacity ± 2 tons applied at 240 Hz. These tests were carried out at load intervals of half a ton and at each load setting four specimens at least were tested

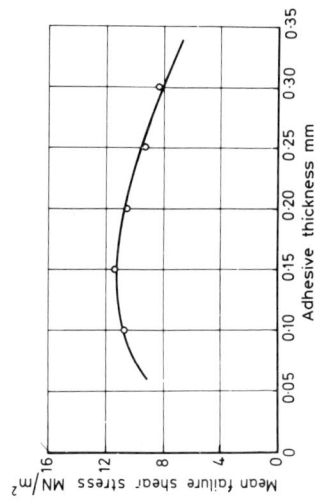

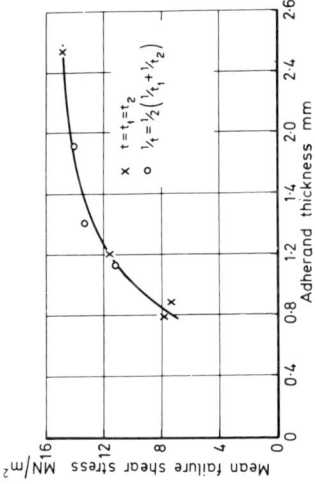

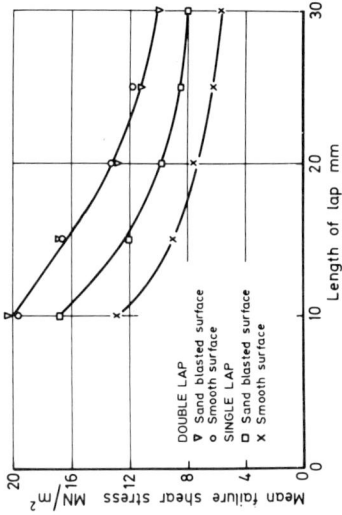

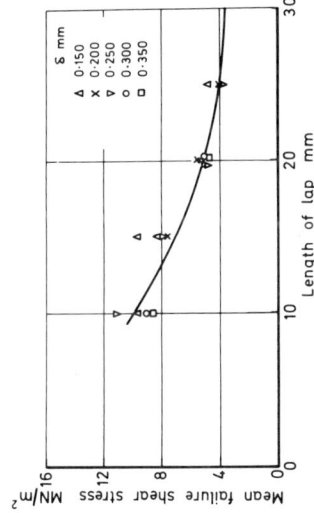

Fig. 8.

Effect of various parameters on mean failure shear stress:
a. adherend thickness, abraded surface l =15 mm, b = 25 mm.
b. adhesive thickness, l = 15 mm, $t_1 = t_2 = 1.55$ mm.
c. lap length.
d. surface roughness.

and the average together with the variance were determined.

For lap joints the fatigue test results were summarised and plotted as a S-N curve in Fig. 9, for three lap joint lengths as given in the figure caption.

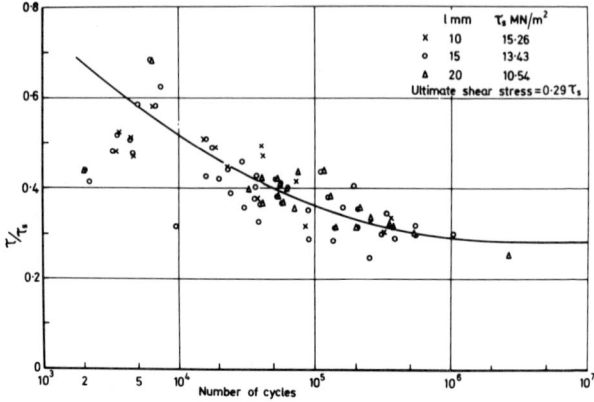

Fig. 9.
Normalised S-N curve for lap joints.

The plot has been normalised by dividing the dynamic load by the corresponding static load for each length. It is noticeable that all the points are scattered around one curve obtained by regression analysis.

For accurate adjustment of the adhesive thickness, two ways were used; one is by inserting two wires the diameter of each equal to the bond line thickness, the second method is by the use of specially designed fixture.

Although the former method is the easiest in the assembly of a joint, it has been envisaged that the wires might cause stress concentration in the bond line which could cause the start of the bondline crack especially in cyclic loading. This phenomenon has been investigated on butt joints in which the bond line thickness was adjusted by each of the above mentioned methods and the result obtained were compared.

In Fig. 10 the S-N curve for each of the adhesive thickness adjustment methods is shown. This corresponds to stress level of $15\mu N/m^2$. The points

corresponding to the mechanical method of thickness adjustment are shown circular
and those for wire insert method are square. This figure shows that the
difference in the S-N curve for both techniques of adhesive thickness control
during the curing process is insignificant.

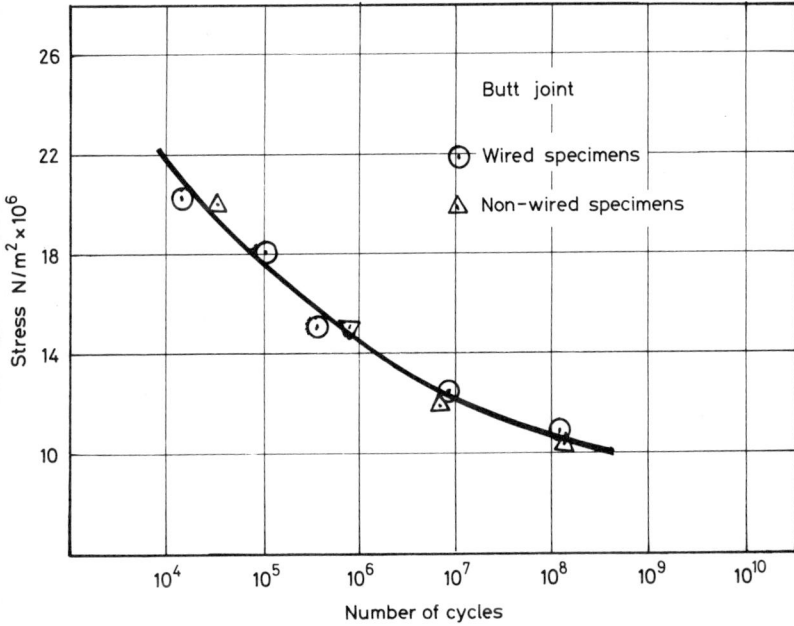

Fig. 10.
S-N curve

However, when investigating the ageing effect of Butt joints assembled
by either of the two thickness control methods, it has been noticed in Fig. 11
that the effect of the control method on the S-N curve is insignificant during
the first week, but after the elapse of the first week, the mechanically
controlled specimen fails after a slightly higher number of cycles than that
corresponding to the wire insert method.

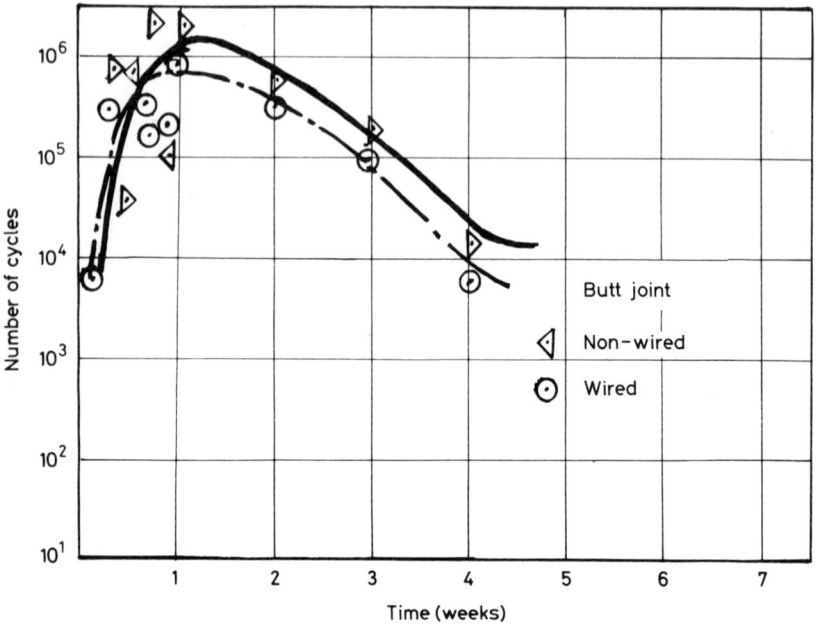

Fig. 11.
Effect of time on Fatigue Life.

It is noticeable from this figure that in the first week the number of cycles at failure increases to a maximum beyond which this number of cycles decreases at a lower rate. It is worth mentioning that this rate of decrease will stabilize after two weeks. At the endurance limit the effect of the joint ageing for a much longer period should be investigated. It is expected, however, that the ageing effect at the endurance limit will be minimal otherwise the bonded horizontal milling machine (2), which has been working for at least eight years, under heavy cutting conditions both in steady state and dynamic loading conditions, would have not lasted over this long period.

2.   Strength of Industrial Specimens

In the fabrication of the bonded horizontal milling machine, the corner

stone of the construction of this machine was the T double containment steel
joints, on which static and environmental investigations were carried out. A
drawing of this T specimen is shown in Fig. 7. In these tests the above mentioned four adhesive bonding agents were used.

The load deflection curves under static loading, for the four types of
adhesive are plotted in Fig. 12 together with that of an equivalent welded T
specimen.

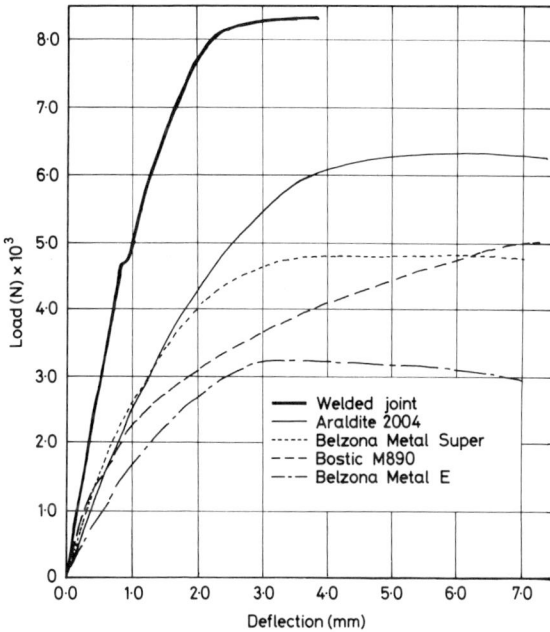

Fig. 12.
Performance of different types of adhesive compared to welding
for a joint under bending load.

It can be noted from Fig. 12 that the highest ultimate load is for the
Araldite 2004 approaching 81% of that of the welded joint followed by that of
Bostic M 890 being 67% of the load of the welded joint. The load at the elastic
limit is highest for the Araldite 2004 being 80% of that of the welded joint,
and the lowest is that of Belzona Metal E, being 33% of the welded joint. The

stiffest joint is that bonded with Bostic M 890 followed by the Belzona metal Super and then Araldite 2004. This shows that the best structural bond is Araldite 2004, since it has the maximum elastic limit and at the same time a low stiffness which permits some relative movement between the mating surfaces and hence higher damping forces.

This justifies the choice of Araldite 2004 in the manufacture of the h horizontal milling machine (4). The main characteristics of the adhesives used in this investigation as obtained from Fig. 12 are summarized in Table VI.

TABLE   VI

| JOINTS | ULTIMATE LOAD $Nx10^3$ | ELASTIC LIMIT | | JOINT STIFFNESS $N/M \times 10^6$ |
		LOAD $Nx10^3$	DEFLECTION $mx10^3$	
Welded	7.65	4.6	0.80	5.75
Araldite 2004	6.20	3.7	1.6	2.31
Belzona Metal Super	4.90	2.5	0.9	2.78
Bostic M 890	5.10	2.0	0.6	3.33
Belzona Metal E	3.30	1.5	0.8	1.86

The effect of ageing on the failure load of the bonded T joints using Araldite 2004 has been investigated over 12 weeks. The results of this study are plotted in Fig. 13 showing that the failure load increases during the first week after which the failure is hardly affected.

One of the most important questions raised by industry interested in the fabrication by bonding   the effect of various environmental conditions on the bonded joint strength. In order to answer this query, various T steel joints bonded by Araldite 2004 were subjected to various harsh environmental conditions. These T joints were left for a period of three months either in open air or immersed in hydraulic oil, kerosine, tap water, coolant or salt water. The average failure loads for these various conditions are tabulated in Table VII. This shows that the joint is slightly affected by the air, hydraulic oil or kerosine. The immersion of the joints in tap water for three months caused a reduction in the failure load by 24% while in coolant results in a reduction of

29% and in salt water 33%. This shows that the environmental surroundings of the joints can in some cases cause deterioration of the joint strength. This however can be avoided by isolating the joint from the environment by some means such as painting which is normally used in cast or welded structures to avoid oxidation.

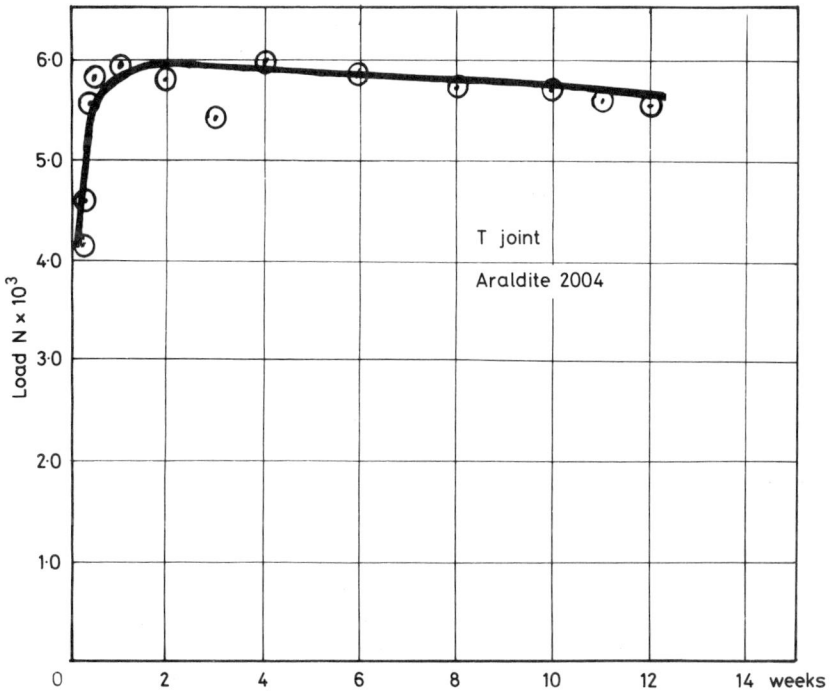

Fig. 13.
Effect of time on joint strength.

TABLE VII

Effect of Environmental Conditions on Failure Loads

ENVIRONMENTAL	AVERAGE FAILURE LOAD $N \times 10^3$	% REDUCTION
Open air	6.2	-
Hydraulic Oil	5.97	4
Kerosene	5.63	9
Tap Water	4.73	24
Coolant Oil	4.43	29
Salt Water	4.13	33

VII   CONCLUSIONS

In this paper it has been shown that fabrication of structures whether small, medium or large, using the bonding technique, can be successfully achieved if a new design philosophy is used. Preliminary work has been carried out in this investigation to answer the questions usually raised by industry. regarding the strength of the bonded joints; effect of harsh environmental conditions, and the type of adhesive suited in each application. The authors are by no means claiming that they established a firm answer to all these questions. They can only give some insight into the problems to be faced in such a complex investigation.

For Araldite 2004, it has been found that the thickness and the lap length has some effect on the joint strength and the surface roughness in single lap joints has a considerable effect, though it has insignificant effects in the case of double lap joints.

The ageing effect on the joint strength is considerable in improving the strength in a week, followed by some reduction up to the period investigated. It has also been shown that the effect of leaving the joint either in air or immersed in kerosine or hydraulic oil has a slight deteriorating effect on the joint strength. On the other hand tap water, coolant, oil or salt water affect the joint strength appreciably. However, the effects of the environment can be minimized if not eliminated, by painting the structure, a process which is used normally in cast of welded structures.

ACKNOWLEDGEMENT

The authors wish to express their thanks to the S.E.R.C. for awarding a large grant to the second author for a large project, a part of which is the work included in this paper. Thanks are also due to G.E.C., Rugby and to Adcock and Shipley, Leicester and last but not least Marwin Cutting Tool Company, Leicester.

Thanks are also due to Mr. P. Thornton for preparing the necessary specimens and assisting in testing various specimens.

REFERENCES

1. Sadek, M.M. "Design for Adhesive Bonding", Engineering, 1979 p. 1290-1292.

2. Showdhury, M.I., Sadek, M.M. and Tobias, S.A. "The Dynamic Character-
    istics of Epoxy Resin Bonded Machine Tool "Structures" Proc. 15th Int.
    M.T.D.R. Conference, MacMillan Press, p. 237-243.

3. Lamb, E.J. and Al-Timimi, K. "Design Concepts for Fabrication of Bonded
    Machine Tool Structures" Proc. 20th Int. M.T.D.R. Conference, MacMillan
    Press 1978, p. 561-567.

4. Sadek, M.M. "Fabrication of Structures Using Bonding by Epoxy Resin
    Adhesives" Part 1, Sheet Metal Industry, p. 837-845. Nov. 1982, Part
    2, March 1983.

5. Chang, H.C., Sadek, M.M. and Tobias, S.A. "Relative Assessment of Dynamic
    Behvaiour and Cutting Performance of a Bonded and Cast Iron Milling
    Machine" Accepted for publication in Journal for Industry, A.S.M.E. 1983.

6. Zang, L.S., Sadek, M.M. and Wale, D.H. "Cutting Performance Comparison of
    Bonded and Brazed End-mills, Presented for publication to A.S.M.E.
    New York.

Chapter 5

ADHESION OF POLYETHYLENE : PHYSICAL AND CHEMICAL ASPECTS

P. DELESCLUSE, J. SCHULTZ and M.E.R. SHANAHAN

Centre de Recherches sur la Physico-Chimie des Surfaces Solides and
Laboratoire de Recherches sur la Physico-Chimie des Interfaces de l'Ecole
Nationale Supérieure de Chimie de Mulhouse
3, rue Alfred Werner, 68093 MULHOUSE CEDEX, France

1 INTRODUCTION

The adhesion of polymers plays an increasingly important role in industrial processes. Apart from the growing use of adhesives for this end, there
are numerous cases in which the assembly of two materials is effected without
the aid of a third component. A good example of this arises in the production
of underground power transmission cables where polymer/polymer adhesion in the
insulation is ensured by simple contact in a process of co-extrusion. The degree of adhesion obtained depends essentially on the materials concerned (composition, molecular weight, etc...) and on the conditions of assembly (temperature, pressure, contact time, etc...).

In a typical power cable, the essential insulating material these days
is a polyethylene, which, in certain cases is crosslinked. The insulator is
sandwiched between two polymeric layers consisting of carbon black filled polyethylenes. These layers serve to avoid the creation of partial electrical discharges between the conductor and the polyethylene insulator and are known
frequently, if erroneously, as "semi-conductors". A potential gradient unavoid-
ably exists at the interface "semi-conductor"/insulator in an operationnal
power cable and in order for this gradient not to lead to a final break-down
of the electrical insulation, good adhesion is necessary. It is known that the
slightest vacuoles, or imperfections of adhesion, at this interface lead to
partial discharges which degrade the insulator and develop with time into potential sites for short-circuits.

The purpose of the present study was thus to consider "semi-conductor"/
insulator adhesion both as a function of its nature and of the conditions of
assembly.

2 MATERIALS AND PRELIMINARIES

Polymer assemblies have been studied in the present work with an aim to understanding the relative contributions to adhesion due to physical and to chemical bonding. Details of the materials employed are as follows.

The polymer representing the insulator in a cable was one of two low-density polyethylene (PE). Both 1020 CN 23 LACQTENE, manufactured by ATO (M_n = 18,500 ; M_w = 115,000 ; melt index 2) and HFDB 4201 S, manufactured by UNION CARBIDE (M_n = 10,000 ; M_w = 40,000 ; melt index 2) were employed. The latter polyethylene contains 20 ppm of antioxidant "Santarox R" and 2 % of dicumyl peroxide (crosslinking agent).

Two materials containing 10 % carbon black (Ketjen EC) were used as "semi-conductors". The first was a polar copolymer of ethylene and ethyl acrylate, manufactured by UNION CARBIDE and this will be referred to as EEA. This copolymer is of melt index 6 and contains 18 % acrylate by weight. The second polymer used was an apolar blend of 80 % polyethylene and 20 % copolymer of ethylene, propylene and a small quantity of diene-monomer. This material will be referred to as EPDM.

The two principal types of manufacturing process used in the cable industry give rise to types of adhesion known by the terms of diffusion and chemical adhesion. In order to study these, three assemblies have been used. The initial assembly can be considered as a model and consisted of PE/PE. The two other assemblies were more representative of industrial conditions and were PE/EEA and PE/EPDM. All three assemblies were in the form of sheets of 4 mm thickness moulded in a heated press, each component being of 2 mm thickness.

The principal test of adhesion employed for all three assemblies was the classic 180° peel test at a peel rate of 5 cm/min. Both peel energy, W_p, and locus of failure were used in the analysis, the latter being studied using techniques such as infra-red spectroscopy, wettability and both optical and electron microscopy.

Measured peel energy depends not only on the intrinsic adhesion of a system, but also on the bulk properties of the constituents and the geometry and temperature of the test. Thus, in order to obtain an approximate order of magnitude of peel energy and to establish reference values, preliminary tests were conducted before embarking upon a more detailed study.

In order to establish a lower limit, pre-crosslinked sheets of polymer were put into contact at 180°C. The interfacial interactions thus produced can be reasonably supposed to be due uniquely to physical bonds of the type Van der Waals and thus the weakest adhesion results.

At the other extreme, it was considered that the highest peel energies available would be those corresponding to the cohesive failure of one of the components. These values were obtained by moulding sheets of double thickness (4 mm) of the various polymers and starting a peel test from a scalpel cut in the centre. The advantage of these two preliminary tests is that the probable maximum and minium peel energies can be obtained without modifying the essential test conditions. Results are summarised in table 1.

Table 1 :

Reference values of peel energy, W_p, for physical (Van der Waals) adhesion and cohesion (peel rate = 5 cm/min)

Assembly	Peel Energy $(kJ.m^{-2})$
Physical Adhesion	
PE(crosslinked)-PE(crosslinked)	2.0
PE " EEA "	1.8
PE " EPDM "	0.5
Cohesion	
PE (crosslinked)	14.5
EEA "	14
EPDM "	15
PE (uncrosslinked)	12
EEA "	14
EPDM "	11

In parallel to peel properties, the dielectric rigidity of the systems in question was studied as a function of the interface. The method is briefly as follows. The insulator/"semi-conductor" is subjected to an increasing applied voltage until breakdown of insulation occurs. The ratio V/d, where V is the voltage at breakdown and d is the total thickness of the polymer assembly, represents the dielectric rigidity.

3 ADHESION BY DIFFUSION

When two uncrosslinked polymers are put in contact either by simultaneous extrusion or by assembly in a heated press, it is possible for diffusion of macromolecules to take place at the interface leading to the creation of an interphase. In these circumstances, physical adhesion results due to Van der

Waals interactions. Nevertheless, the degree of adhesion should be relatively high since a very large number of interactions is possible between the diffused species. This mode of adhesion by diffusion was proposed by Voyutskii[1] and is generally known by the term autohesion when the two polymers in contact are identical. However, when the constituents are different, the question is often asked as to whether adhesion can be attained for two reasons. Firstly, there must be mutual compatibility between the materials in contact or sufficient diffusion will not take place on thermodynamic grounds. Secondly, the relative mobility of interdiffusing species must be sufficiently great under the given conditions of temperature, pressure and contact time.

In the present study, it is reasonable to suppose that the first criterion is satisfied since both the insulator and the "semi-conductor" contain very high proportions of the same fundamental material - polyethylene. Concerning the second criterion, this has been studied experimentally. It was found that the overriding parameter governing uncrosslinked insulator/"semi-conductor" adhesion was the temperature. Figure 1 shows the evolution of the logarithm of peel energy, W_p, vs. the inverse of absolute temperature, T, of assembly for the two systems PE/EEA and PE/EPDM. This figure demonstrates that for an assembly temperature $\lesssim 80°C (1/_T \gtrsim 2.8 \times 10^{-3} K^{-1})$, resulting adhesion is poor but above this temperature up to ca. 120°C ($1/T \approx 2.55 \times 10^{-3} K^{-1}$), adhesion increases rapidly. Above this last temperature up to 180°C, the evolution is relatively slow. This evidence alone, however, does not constitute proof of adhesion due to interdiffusion. It is conceivable that higher temperatures simply lead to a better direct contact and therefore adhesion of the materials due to softening and/or melting. Nevertheless, fig. 1 suggests that two different processes are involved, one from ca. 80 to 120°C and the other from ca. 120 to 180°C. In fact 120°C corresponds approximately to the melting point of the polymers in question, and these results, comparable to those found for the autohesion of certain polyethylenes[2] can be explained by the destruction of the crystalline structure. Below the melting point, adhesion is limited by the rate of decomposition of the morphological structure of the polyethylene. Crystalline zones (spherulites) must break up into lamellae, or monocrystals of polyethylene, before large numbers of individual macromolecular chains become mobile. This process requires a high activation energy. Assuming that the measured adhesion, W_p, is directly related to an activated process such that

$$W_p = K \exp [- E_a/RT] \qquad \qquad \ldots \qquad 1.$$

where K is a constant, E_a is an activation energy for the "creation" of adhesion (or molecular mobility), R is the gas constant and T the absolute temperature, the gradient of the upper portion of fig. 1 (corresponding to $80°C \lesssim T \lesssim 120°C$) leads to $E_a \simeq 110$ kJ.mole^{-1}. During the said decomposition, the only macromolecular chains capable of diffusion are those either being freed from the crystalline structure or originating from the amorphous interspherulitic regions.

In contrast, above the melting point of ca. 120°C, all the macromolecules are free to diffuse and a reduced apparent activation energy will result in eq. 1. Analysis of this lower portion of fig. 1 ($120°C \lesssim T \lesssim 180°C$) leads to $E_a \simeq 4$ kJ.mole^{-1}.

In addition to the direct study of peel energy, the loci of failure were examined. The general result observed was that failure was cohesive within the "semi-conductor" and that the depth at which failure occurred increased with the assembly temperature. This result was obtained from a study of failure surfaces using a combination of optical microscopy, infra-red transmission spectroscopy and wettability techniques. In the case of optical microscopy, it was noted that large numbers of carbon particles were present, after failure, on the insulator. Infra-red spectroscopy was used only in the case of the PE/EEA assembly where it was noted that the C = O peak in the polyethylene increased with the assembly temperature, as shown in table 2.

Table 2 :

I.R. transmission data of insulator
after failure (PE/EEA, diffusion)

Assembly temperature (°C)	80	90	100	110	120	150	180
C=O peak intensity at 1720 cm^{-1} (arbitrary units)	1	2	6	11	13	19	23

Supporting evidence from wettability measurements was based on exploitation of Wenzel's roughness factor[3] :

$$r = \frac{\cos \theta}{\cos \theta_o} \qquad \qquad \text{...} \qquad 2.$$

where θ and θ_o are the contact angles of water measured respectively on the peeled surface and on a reference surface of the "semi-conductor". In physical terms, r represents the ratio of the real interfacial area polymer/water and the apparent geometrical area. The similarity of the values of r found for the insulator and "semi-conductor" surfaces after separation shown in table 3 tends to reinforce the belief that failure occurred in a surface layer of the latter.

Table 3 :

Roughness data of the failure surfaces
of systems PE/EEA and PE/EPDM (diffusion)

Assembly Temperature (°C)	PE/EEA $r = \dfrac{\cos \theta}{\cos \theta_o}$ (EEA)		PE/EPDM $r = \dfrac{\cos \theta}{\cos \theta_o}$ (EPDM)	
	PE	EEA	PE	EPDM
80	1	1	1	1
90	3	3	1	2
100	5	3	4	5
110	7	5	9	9
120	6	5	9	9
150	7	6	9	9
180	6	6	9	9

Both surface rugosity and peel energy increase in a similar manner with assembly temperature -a fact suggesting that peel energy is directly related to the energy of cohesion of the "semi-conductor" surface layer. Both rugosity and peel energy increase rapidly for assembly temperatures from 80 to ca. 120°C but above 120°C, evolution is much slower. This evolution can only result from increased interdiffusion. Adhesion due simply to strictly interfacial Van der Waals bonding could not reasonably be expected to affect the rugosity of failure surfaces to such a degree.

As far as electrical properties are concerned, it was noticed that the dielectric rigidity of the assemblies decreased with increasing assembly temperature ; this being particularly the case for the system PE/EEA (table 4). This reduction was attributed to diffusion of polar species from the "semi-conductor" into the polyethylene. Verification of this hypothesis was obtained by incorporating EEA in PE (1 : 10) and examining two electrical properties, as resumed in table 5. It can be seen that the inclusion of EEA decreases the dielectric rigidity markedly. The second test amounts to a simple measurement of time to electrical breakdown of an insulating polymer between a flat and a point electrode at high potential, and is described in ref. 4. Here it can be

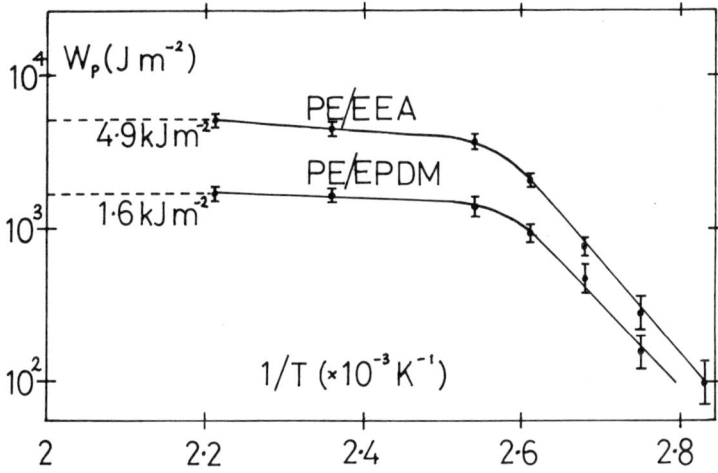

Fig. 1.   Peel energy, W_p, vs. reciprocal assembly temperature, 1/T, for uncrosslinked systems.

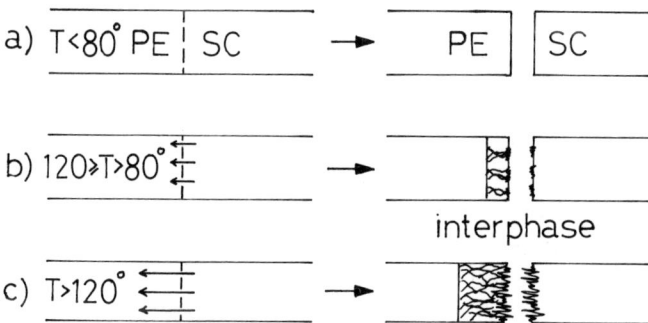

Fig. 2.   Model of diffusion adhesion mechanism depending on assembly temperature.

Table 4 :

Dielectric rigidity as a function of assembly
temperature for a system PE/EEA (diffusion)

Assembly Temperature (°C)	120	140	180
Dielectric Rigidity (kV/mm)	55	50.5	47.5

Table 5 :

Electrical properties of PE and
PE containing 10 % EEA

	Dielectric Rigidity (kV/mn)	Time to Electrical Breakdown under High Voltage (hr)[4]
PE	71	150
PE + 10 % EEA	60	20

seen that inclusion of EEA provokes a drastic reduction in effective insulator
life.

The above results lead to the suggestion of the following model explai-
ning the present form of adhesion due to diffusion. Line a) of fig. 2 represents schematically the process of adhesion at an assembly temperature < 80°C.
No diffusion takes place due to very low molecular mobility and adhesion thus
depends on interfacial Van der Waals bonding. The result is very poor adhesion.
In line b), representing assembly at 80 to 120°C, it is suggested that diffusion of low molecular weight species takes place from the "semi-conductor" into
the insulator, but that this effect will be limited essentially to the amorphous zones of the former. The result is stronger adhesion due to the much increased number of physical bonds formed during diffusion. Failure takes place
in a weak boundary-layer on the "semi-conductor" at the limit of the interphase thus produced. Above 120°C, line c), diffusion is much enhanced since
higher molecular weight species are liberated from the now melted crystalline
structure. The overall effect is an increase in the energy of cohesion of the
surface layer of the "semi-conductor", and a corresponding increase in measured
peel strength.

The existance of interphases as suggested in fig. 2, lines b) and c) has
not been successfully demonstrated in the present study. Nevertheless, Yamakawa

showed the existance of interphases of a few microns thickness in not dissimilar assemblies, by employing microinterferometry[5].

4 CHEMICAL ADHESION

Adhesion of a chemical nature is the form most commonly exploited in the
manufacture of high tension electric cables. Industrially, assembly is effected in two stages ; the first stage being extrusion of the insulating polymer
and the "semi-conductor" together. The materials employed contain 2 % of dicumyl peroxide (DCP) and thus the second stage, involving a thermal treatment by
steam under pressure, leads to crosslinking. This crosslinking confers on the
materials an improvement in mechanical properties, especially creep resistance.
The latter is of the utmost importance since high tension cables are often
subjected to working temperatures of 70°C which can, in the case of a short-
circuit, easily rise to 150°C.

Although in practice the above process is very successful, the mechanism
of adhesion is poorly understood and it is for this reason that the present
study was undertaken in order to elucidate the role played by crosslinking.

Since measured peel energy depends on various test parameters such as
geometry and bulk properties of the materials in question, as well as intrinsic adhesion[6,7,8], it was necessary to use a technique allowing only this last
factor to change to any appreciable degree. This was done as follows using a
method suggested by Gent[9,10]. The various polymers, containing 2 % DCP by
weight, were molded into sheets at 120°C (the insulator used already contained
DCP, for the "semi-conductors" the DCP was incorporated). At this temperature,
the rate of decomposition of DCP is negligible. The preparation of the final
assemblies was then effected in two steps. In the first step, partial pre-
crosslinking of the still separate sheets of polymer was achieved by heating
to 140°C. The decomposition of DCP follows first order kinetics and the degree
of crosslinking will be proportional to the fraction of DCP decomposed ($\frac{c}{c_o}$).
This latter depends on the time, t, at the said temperature, following
the relation :

$$\frac{c}{c_o} = [1 - \exp{(- k_d t)}] \qquad \qquad \dots \quad 3$$

with $k_d = \ln{2/t_{1/2}}$ and $t_{1/2} = 32$ minutes at 140°C[11]. After the desired time
at 140°C, rapid cooling to 120°C, followed by slow cooling to ambient temperature ensued. The second step of the operation involved assembly of the two
components and heating at 180°C under pressure for 10 minutes. This ensured

the decomposition of the remaining DCP $(1 - c/c_o)$. Crosslinking both in the bulk of the phases and at the interface takes place in this second step. The complete procedure is summarised in fig. 3.

The advantage of this technique is that at the end of the operation, crosslinking is complete in the bulk of the assembly and thus the mechanical properties are unaffected, whatever be the nature or degree of interfacial bonding. This being the case, measured peel energy only varies as a function of the reversible energy of adhesion (W_o), all other parameters being approximately constant. The measured peel energy is thus a direct function of the resistance of the interface, even if it is difficult a priori to find a quantitative relationship between the two.

Figure 4 shown the results of measured peel energy as a function of the percentage of DCP decomposed during assembly of the two components $(1 -c/c_o)$, in the case of the model system PE/PE. At point A, the two sheets were entirely crosslinked before assembly. The resulting adhesion was thus uniquely physical and failure surfaces, being totally smooth, confirmed the existence of poor adhesion. Certain sections of some assemblies, even when crosslinked, displayed this phenomenon (line AZ in fig. 4). This was attributed to the existence of non-negligible temperature gradients near the edges of the assembly press. These few cases are therefore not considered to be representative. Following the curve AB, it is clear that W_p increases with the percentage of DCP decomposed during assembly. Higher values of W_p corresponded to "mixed" failure surfaces in which the propagation of failure took place near the interface. These surfaces were very rough and inhomogeneous -their irregularity increasing with peel energy. Finally for the section BC in fig. 4 in which most of the DCP was decomposed during assembly, true peel was not observed. Instead, failure tended to propagate in the bulk of one or other of the two components. Under these conditions of cohesive failure, the energy of adhesion is equal to or greater than the cohesive energy of the crosslinked PE. The value of 15 kJ.m^{-2} obtained for W_p is indeed very close to that found in the test of cohesion (table 1).

Figures 5 and 6 represent equivalent data for the assemblies PE/EEA and PE/EPDM. The basic relationship is similar. However, in both of these cases, there is an important change in gradient at values of $(1 - c/c_o)$ of ca. 20 %. This has been attributed to a transition from elastic to plastic behaviour of the "semi-conductor" as peel energy increases[4].

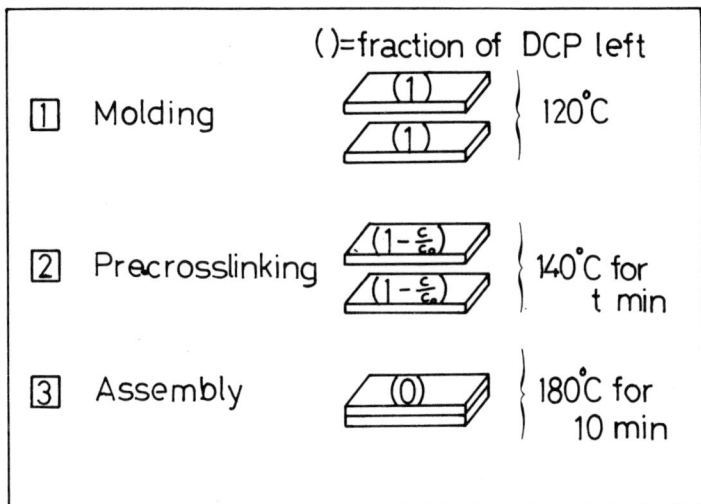

Fig. 3. Preparation of assemblies for the study of chemical adhesion.

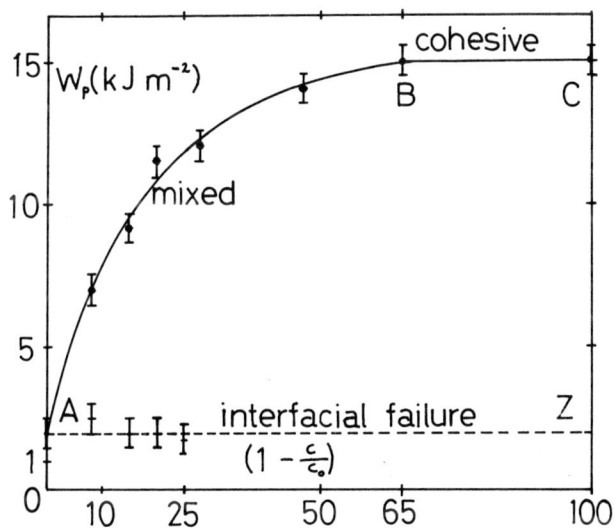

Fig. 4. Peel energy, W_p, vs. DCP concentration of assembly, $(1 - \frac{1}{C_o})$ for system PE/PE.

The overall picture is that the resistance to peel of these systems increases with the quantity of DCP decomposed during assembly. Clearly the DCP plays an important role in adhesion and the next section tries to elucidate this role.

4.1 Role of the DCP

The crosslinking of PE by DCP is very efficient and only slightly dependent on the quantity of the latter present[12,13]. A suggested crosslinking scheme based essentially on the combination of macroradicals[14,15] is shown belo⎵

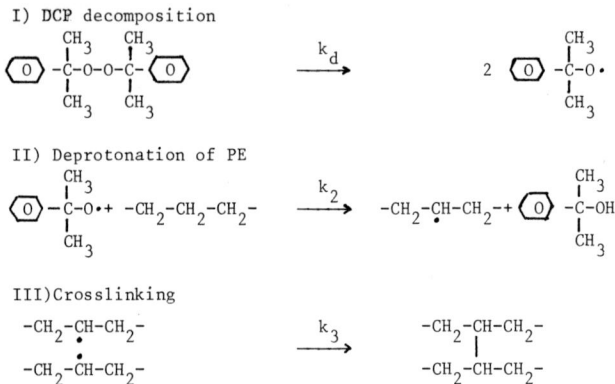

I) DCP decomposition

II) Deprotonation of PE

III) Crosslinking

The slowest and therefore rate determining step is the decomposition of the DCP, which obeys first-order kinetics.

As a result, there are two possible mechanisms which could reasonably explain the observed results. The first possibility is that adhesion is due to a phenomenon of interdiffusion of macromolecular chains. This interdiffusion is limited by reduced molecular mobility in the case of a polymer already partially crosslinked before assembly. The second possibility is that adhesion could be attributed to the formation of covalent bonds at the interface of the assembly. This process would involve radicals and would therefore be akin to crosslinking. In this case, internal and interfacial crosslinking would be simultaneous and the term co-crosslinking is thus adopted for the latter. The existence of such interfacial covalent bonding was shown using the following technique in which molecular diffusion is extremely limited. Two sheets of totally crosslinked PE were painted with melted DCP at 80°C (ca. 1 mg.cm^{-2}), assembled and then heated at 180°C for 10 minutes under pressure. The peel energy of such a system was found to be 9 kJ.m^{-2} compared to 2 kJ.m^{-2} for the same assembly not having undergone the application of DCP. Equivalent results found for the assembly PE/EPDM were respectively 5 and 0.5 kJ.m^{-2}. In addition, the fracture surfaces of the DCP painted samples were very similar to those having undergone co-crosslinking.

Although these results clearly show the existence of covalent bonds at the interface due to the decomposition of DCP, the mechanism of macromolecular diffusion cannot be rejected since the formation of chemical bonds can only take place after a certain degree of diffusion.

A theoretical attempt at assessing the relative importance of chemical bonding and of diffusion is a forbidding task and, as such, experimental observations were relied upon as indicating that the latter is only a secondary factor in adhesion at the interface. Comparing fig. 5 and 6 with fig. 1, it can be seen that the maximum peel energy in the case of chemical adhesion is far greater than that for adhesion by diffusion. Nevertheless, any contribution due to diffusion in fig. 5 and 6 will be relatively reduced with respect to that in fig. 1 since in the former cases, crosslinked systems are being considered whereas in the latter, no crosslinking takes place. In addition, surface roughness after separation, although existent in the case of adhesion by diffusion, is relatively superficial, in constrast to very marked surface rugosity in the case of chemical adhesion. The above, combined with the fact that the cohesive strength of the materials in question is only slightly affected by crosslinking (table 1) suggests that the differences in peel resistance must be due to different adhesion mechanisms. Lastly, it can be seen from table 6 that the dielectric rigidity in the case of chemical adhesion is actually greater than that found for physical adhesion, whereas adhesion simply by diffusion lowers dielectric rigidity. This fact seems incompatible with any suggestion that chemical adhesion is of minor importance compared to diffusion.

Table 6 :
Dielectric rigidity in kV.mm^{-1} for systems
assembled at 180°C for 20 mins

	Chemical Adhesion	Physical Adhesion	Adhesion by Diffusion
PE/EEA	64	60	47
PE/EPDM	74	67	60

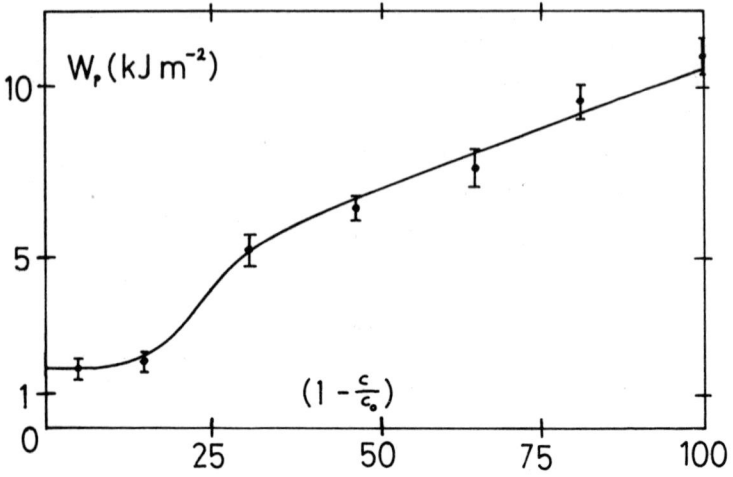

Fig. 5. W_p vs. $(1 - \frac{1}{C_o})$ for system PE/EEA.

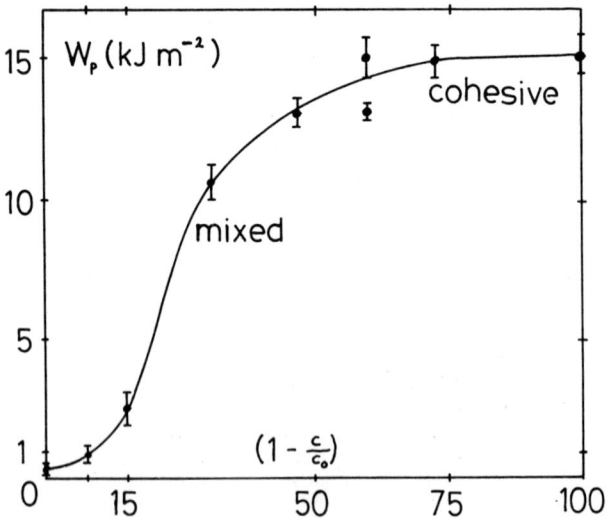

Fig. 6. W_p vs. $(1 - \frac{1}{C_o})$ for system PE/EPDM.

4.2 Number and Nature of Chemical Bonds

In the case of the model assembly PE/PE, an attempt was made to estimate the number of chemical bonds developed at the interface. Point C in fig. 4 represents total DCP decomposition during assembly. Since failure is cohesive in this case, it is reasonable to suppose that the number of bonds created per unit area of interface is comparable to the number of elementary molecular chains crossing unit area of surface taken arbitrarily within the bulk of the crosslinked PE. The term "elementary chain" refers to a section between two points of crosslinking. The symbol ν is now introduced to represent the number of such chains per unit volume of PE partially crosslinked before assembly and ν^{100} in the case of total crosslinking. Both ν and ν^{100} were obtained from swelling measurements in boiling xylene and using the Flory-Rehner theory[16]. Assuming a homogeneous distribution, the number of chains crossing unit area in the bulk, ν_s, will be given by $\nu^{2/3}$. As such, the number of interfacial bonds, ν_i, will be given approximately by $\nu_i = \nu_s^{100} - \nu_s$. Using this simple theory and plotting peel energy, W_p, vs. ν_i leads to the results shown in fig. 7. It can be seen that the relation is essentially linear, as might be expected intuitively. Of fundamental importance is the fact that this shows that the energy dissipated in a peel test, W_D, involving plastic deformation is directly proportional to the reversible energy of adhesion (cohesion), W_o.

$$W_p \simeq W_D \simeq \text{constant} \times W_o \qquad \qquad \ldots \qquad 4$$

This relation is similar to that obtained by Gent and Schultz for viscoelastic materials[6].

Cohesive failure occurred in the assembly PE/PE for values of $(1 - c/c_o)$ greater than ca. 65 %. Since molecular entanglements should exist within the bulk of the PE, but not at the interface of the assembly, it may be supposed that the "effective" degree of interfacial crosslinking (by contrast to the "true" degree involving chemical bonding) should be smaller than that in the bulk. As such, some type of "supplementary" bonding should be present at the interface to counteract this deficiency and thus explain the observed cohesive failure. The following mechanism is proposed in which a certain number of peroxide bonds are formed with the participation of atmospheric oxygen.

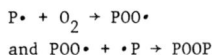

$P\cdot + O_2 \rightarrow POO\cdot$

and $POO\cdot + \cdot P \rightarrow POOP$

where P represents a PE chain.

Clearly the direct detection of a relatively small number of such bonds
presents great difficulty and as such, the majority of methods attempted
proved fruitless (differential I.R. spectroscopy, desorption of oxygen before
assembly, thermal treatment). Nevertheless, a positive result was obtained by
chemical reduction in boiling aniline. This treatment is efficient for the
destruction of peroxide bonds and depends on the following mechanism[17].

$$2C_6 H_5 NH_2 + 2POOP \rightarrow C_6 H_5 N = NC_6 H_5 + 4POH$$

Experimentally it was observed that the above treatment led to a marked
decrease in peel energy as shown in fig. 7. It is nevertheless true that the
abrupt decrease in peel resistance is not due entirely to the destruction of
peroxide bonds. The PE is plastified to some extent by aniline lowering its
cohesive energy from ca. 15 to 9 kJ.m^{-2} after 10 hours treatment. However, as
can be seen in fig. 8, the equivalent decrease in peel energy for the PE/PE
assembly is from 15 to 5 kJ.m^{-2}. This difference can only reasonably be explained by the destruction of peroxide bonds in the second case. In addition,
whereas failure is cohesive before aniline treatment, after treatment it can
be seen that failure is truly interfacial.

5 CONCLUSIONS

Polymer adhesion has been studied with particular reference to materials
used in power transmission cables. Two processes are employed industrially and
the adhesion resulting from each has been studied.

In the first part of the study, assemblies consisting of crosslinked polymers have been shown to depend essentially on diffusion properties for adhesion. Macromolecular chains diffuse mainly from the "semi-conductor" into the
insulator leading to adhesion of a physical nature between neighbouring chains.
Resulting peel resistance is fair but, by contrast, the electrical properties
of the assembly measured as a function of dielectric rigidity, are very feeble.

The second part of the study involves chemical adhesion between crosslinked polymers. Adhesion increases markedly with the quantity of DCP decomposed during assembly. This phenomenon has been attributed to the formation of
covalent bonds at the interface depending on a mechanism initiated by the decomposition of the DCP. In the case of the model assembly PE/PE, the number
of covalent bonds at the interface has been estimated, leading to the result
that adhesion is directly proportional to this number. The existence of a
small proportion of peroxide bonds at the interface has been shown indirectly.

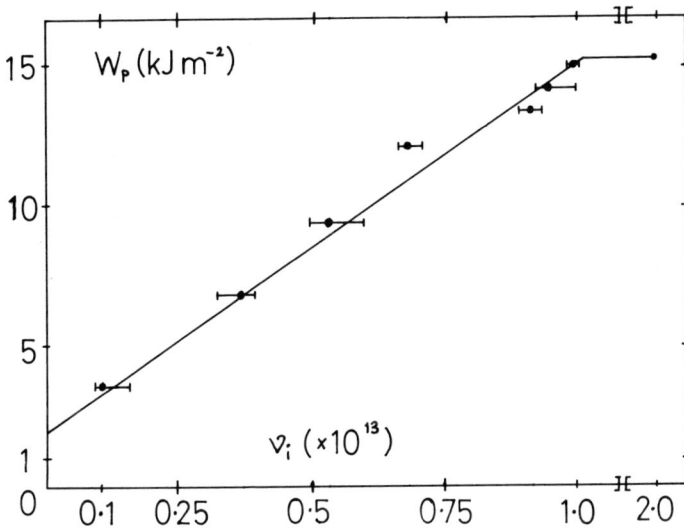

Fig. 7. W_p vs. number of interfacial bonds / cm^2, ν_i, for assembly.

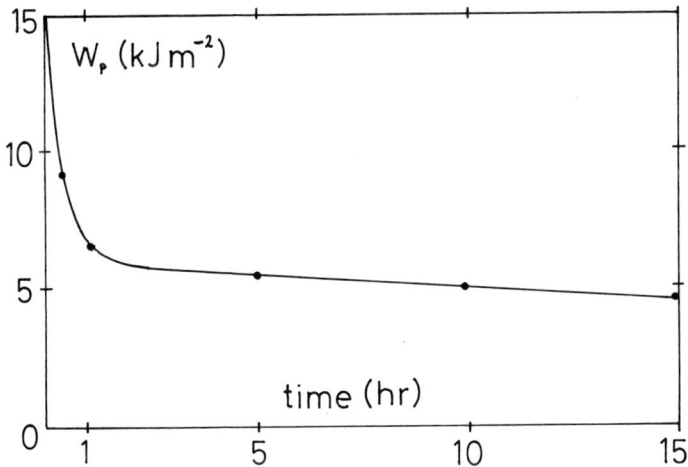

Fig. 8. W_p vs. time of treatment in boiling aniline of totally co-crosslinked PE/PE assembly.

Not only is the mechanical resistance very good when chemical adhesion is employed, but also the electrical properties improve.

An overall conclusion is that in the construction of power cables, chemical adhesion between polymers is clearly superior to physical (diffusion) adhesion.

REFERENCES

1. Voyutskii, S.S.,'Autohesion and Adhesion of High Polymers', Interscience, New York, 1963

2. Shtarkman,B.P., Voyutskii, S.S., Kargin, V.A., Vysokomol. Soyed., 7 (1), 135, 1965

3. Wenzel, R.N.,Ind. Eng. Chem., 28, 988, 1936

4. Delescluse, P.,Thèse de Docteur-Ingénieur, Université de Haute-Alsace, 1980

5. Yamakawa, S.,Polym. Eng. Sci., 16, 6, 1976

6. Gent, A.N., Schultz, J.,J. Adhesion, 3, 281, 1972

7. Andrews, E.H., Kinloch, A.J.,Proc. Roy. Soc., London, A 332, 385, 1973

8. Kendall, K.,J. Adhesion, 5, 179, 1973

9. Gent, A.N., Private communication, 1977

10. Gent, A.N., Forum International Adhesion, Lyons, Sept. 1979

11. Brandup, J., Immergut, E.H., 'Polymer Handbook', Interscience, New York, 1967

12. Miller, A.A., J. Polym. Sci., 42, 441 1960

13. Loan, L.D., Rubber Chem. Tech., 40, 163, 1967

14. Rheinfeld, D., Gummi Asbest Kunst., 7, 80, 1975

15. Chow, Y.W., Knight, G.T., Ibid, 31, 716, 1978

16. Flory, P.J., Rehner, J., J. Chem. Phys., 11, 512, 1943

17. Wheland, G.W., 'Advanced Organic Chemistry', Interscience, New York, 1949.

NOTATION

E_a	activation energy for molecular mobility
R	gas constant
T	temperature
V	voltage
W_o	reversible energy of adhesion
W_p	peel energy
c	concentration of DCP
d	thickness of assembly
k	rate constant
r	Wenzel's roughness factor
t	time
θ	contact angle
ν	number of elementary chains/unit volume (or surface).

Chapter 6

THE EFFECT OF SINGLE FUNCTIONAL GROUPS ON THE ADHESION CHARACTERISTICS OF
POLYETHYLENE

Alison Chew*, D M Brewis,* D Briggs[+] and R H Dahm*

* School of Chemistry, Leicester Polytechnic, PO Box 143,
  Leicester, UK
[+]ICI plc, Petrochemicals and Plastics Division, PO Box 90,
Wilton, Middlesbrough, UK

1 INTRODUCTION

In order to enable a polyolefin to be bonded, coated or printed upon,
it is usually necessary to improve its adhesion by pretreating the surface.
Effective pretreatments have been available commercially for over 30 years,
in particular corona discharge treatments, flame treatments and chromic acid
etching[1]. These are used to treat films, thicker sections such as bottles
and small or irregularly shaped objects respectively. There has been much
controversy regarding the reason why these pretreatments improve the adhesion
of polyolefins[2].

Polyethylene (PE) and polypropylene (PP) have low surface energies
(γc, critical surface tension of wetting $\simeq 31$ mN m^{-1}) and it has been suggested
that the surface treatments are effective because of an increase in the
surface energy, ie. the wettability of the surface is improved[2].

A second explanation is that the poor adhesion of polyolefins is due
to the presence of weak boundary layers and the function of a surface pre-
treatment is the removal of such material. In addition morphological changes
occur when polyolefins are commercially pretreated. It is therefore possible
that mechanical keying plays an important part in improving the adhesion of
PE and PP.

Alternatively, the introduction of specific functional groups onto the
polymer surface may be responsible for the large improvement in adhesion,
because of a large increase in the interaction across the interface.

The commercial pretreatments all introduce a complex variety of functional groups into the surface. The groups introduced by various pretreatments include carbonyl, carboxylic acid, hydroxyl, hydroperoxides, amino and sulphonic acids. However, it has rarely been possible to determine which individual functional groups are responsible for the increases in adhesion that are observed when commercial pretreatments are used.

For this reason, work has been carried out to introduce individual functional groups, one at a time, into polyolefin surfaces, followed by a study of the effect of each group on the adhesion to an epoxide adhesive. This has been achieved by brominating the surface followed by a series of synthetic reactions to introduce individual groups into the surface. These functional groups have then been derivatised with reagents that react specifically with one functional group, thereby introducing a heteroatom 'label' that may be readily detected by X-ray photoelectron spectroscopy (XPS)[3]. Thus individual oxygenated groups may then be identified unambiguously.

The modifications to the polyethylene surface have also been assessed using multiple internal reflectance infrared spectroscopy (MIR), contact angle measurements and the destructive testing of composite butt joints.

However, problems arise when carrying out reactions at polymer surfaces as compared to organic solution chemistry. The reactions must proceed in high yields as it is not possible to purify or separate the products. Heterogeneous catalysts cannot be used as the brominated polymer is never in solution. Although the stereochemical features of the homogenous reactions are fairly well understood, little is known about the stereochemical requirements of the surface reactions. The steric requirements may be facilitated by the use of swelling solvents but such solvents may cause migration of functional groups into the bulk or low molecular weight material migrating to the surface.

2 EXPERIMENTAL

2.1 Materials

Additive free low density polyethylene (LDPE) film (0.15 mm thick) was supplied by Petrochemicals and Plastics Division, ICI.plc. This was Alkathene 33. Additive free high density polyethylene (HDPE) was supplied by BP Chemicals plc and was type Rigidex 002 55. This was pressed against poly(ethylene terephthalate) film which had previously been extracted with carbon tetrachloride.

All solvents that were used were rigorously dried before use by the methods of Perrin, Armarego and Perrin[4]. White spot nitrogen was further purified by passing it through a column of molecular sieves (Linde type 5A) and over an oxygen scavenger BASF catalyst R3-11. This is a pelleted form of finely divided copper on an inert support. It is claimed that this method reduces the water and oxygen content to less than 1 ppm[5].

2.2 Bromination

Polyethylene was brominated by exposing the polymer film to saturated bromine vapour in a nitrogen atmosphere with simultaneous irradiation by tungsten light. Two 150 W light bulbs were used, one on either side of the bromination vessel, 400 mm distant from the film.

2.3 Reduction of the Brominated Surface

The brominated polymer was reduced by refluxing in 2-methyl tetrahydrofuran (200 ml) containing tri-n-butyl tin hydride (10 g) and a thermal initiator (Azo-iso-butyronitrile 0.1 g) for 6 hours.

2.4 Dehydro Bromination

It was possible to dehydrobrominate the brominated polyethylene by immersing in THF/n-butylamine (50:50 by volume) for one week.

2.5 Reactions of the Unsaturated Sites

The resulting unsaturated polymer was then subjected to the following reaction scheme:-

a) Hydroboration-oxidation: this reaction was carried out according to the method of Vogel[6] in order to introduce a single hydroxyl group for each double bond onto the surface of polyethylene.

b) Potassium permanganate oxidation: the unsaturated polyethylene film was immersed in a dilute alkaline aqueous solution of potassium permanganate (2×10^{-2} molar) for 10 minutes. This treatment introduces vicinal hydroxyl groups into the surface.

c) Permanganate periodate oxidation: an aqueous solution containing potassium permanganate (5×10^{-4} molar) and potassium iodate (4×10^{-2} molar) was prepared and unsaturated polyethylene was immersed in it, in the dark, for 20 hours at room temperature. This reaction gives rise to carboxylic acid groups in the polyethylene surface. When untreated polyethylene was reacted with each of the above reagents no appreciable amount of oxygen was introduced, as shown by XPS.

2.6 Reaction of Brominated Polyethylene

Brominated polyethylene was treated in a solution of silver tetrafluoroborate in dimethylsulphoxide for 4 days. Triethylamine was then added and
the reaction left for a further three days, in order to introduce keto groups
into the surface of the polymer. The functional groups that were introduced
into the polyethylene surfaces were derivatised with the following reagents:-

    i)   Pentafluorophenylhydrazine (PFPH)[3]

    ii) Di-isopropoxytitanium bisacetylacetonate (TAA)[3]

    iii) Thallium ethoxide (TlOEt)[7]

These reagents introduce heteroatom labels that may be readily identified by
XPS.

2.7 Contact Angles

A photographic method was used to measure the contact angles of water
on the surface of polyethylene. The apparatus consisted of the following:-

    a tungsten light source

    a diffuser

    a 10 cm convex lens to give a parallel beam of light

    a monochromator

    a table on which the sample is placed and covered with a Perspex cube

    an Agla syringe (to dispense drops of 10 µl)

    a Practica LT3 camera, together with a reversing ring and extension
    rings (Asahi Opt. Co. Nos 1, 11 and 111)

The whole apparatus was assembled on an optical bench. The film was developed and the negatives were projected onto a screen. The contact angle
was measured either directly or by measuring the width and height of the
drop and applying the equation:-

$$\tan \frac{\theta}{2} = \frac{2h}{d}$$

This work is being extended to include other liquids, eg. glycerol, formamide and methylene iodide. This will then enable the polar and dispersion
contributions to the work of adhesion to be determined.

2.8 X-Ray Photoelectron Spectroscopy

All XPS data was obtained at Petrochemicals and Plastics Division,
ICI Plc, using an AEI ES200B spectrometer employing MgKα exciting radiation
(1253.6 eV).[8]

2.9 Multiple Internal Reflectance IR

MIR spectra were run on a Perkin Elmer 683 infrared spectrometer together with a SpecAc 25 reflection unit. The equipment was purged with oxygen free nitrogen at all times, in order to eliminate atmospheric water and carbon dioxide. The internal reflection crystal used was a thallium bromide/thallium iodide mixed crystal (KRS-5). Prior to each spectrum being run, the crystal was checked for contamination by using a "paste holder" which allows a spectrum of the blank crystal to be run.

2.10 Preparation and Testing of Bonds

Mild steel butts of 28 mm diameter were pretreated by abrading on a surface grinder and washing in methyl ethyl ketone. 32 mm diameter circles of polyethylene film were sandwiched between two vertical butts. The butts were kept in position by means of a special jig. The adhesive used was Epikote 828, which is an epoxy resin based on Bisphenol A, together with an amine hardener, Synolide 926. These were used in a ratio of 5:3 by weight and the adhesive was cured for 16 hours at 323 K. The joints were tested on a Monsanto 2000 tensometer at a crosshead speed of 49 mm per minute.

3 RESULTS AND DISCUSSION

The joint strength of untreated HDPE is less than half that of LDPE. This is possibly due to a very low level of oxidation and unsaturation in HDPE. The amount of unsaturation was determined by exposing the films to bromine vapour in the dark. The XPS results may be seen in Table 1. Extraction experiments show that HDPE contains very little low molecular weight material in comparison with LDPE although this would tend to improve the adhesion.

3.1 Bromination of Polyethylene

It has been possible to improve the adhesion of both LDPE and HDPE to an epoxide adhesive by brominating the surface. In order to achieve the same increase in joint strength for HDPE as for LDPE, it was necessary to use considerably longer irradiation times. This is very probably because HDPE contains fewer reactive positions, eg. branch points or allylic positions, than LDPE. Adhesion levels and XPS data are presented in Table 1.

The bromination of polyethylene is a reproducible process, it being possible to produce films with known amounts of bromine and small quantities of oxygen (<1%). This oxygen level may only be maintained if purified nitrogen is used to thoroughly flush the system. If oxygen is allowed to enter the brominating vessel, other groups are introduced and subsequent interpretation of results is complicated.

Table 1 - Bromination of Polyethylene

	Joint Strength MN m^{-2} ± s dev	Type of Failure	XPS Data/Atomic ratios C:Br	C:O
Untreated LDPE	1.4 ± 11%	I	–	196:1
Untreated HDPE	1.77 ± 20%	I	–	333:1
Brominated LDPE (1 hour irradiation)	7.5 ± 14%	I	22:1	80:1
Brominated HDPE (2½ hours irradiation)	5.6 ± 17%	I	19:1	110:1
Brominated LDPE in dark (1½ hours)	(not known)		146:1	142:1
Brominated HDPE in dark (3 hours)	(not known)		539:1	149:1

I = apparent interfacial failure

3.2 Reduction of Brominated Polyethylene

To elucidate the importance of weak boundary layers, attempts have been made to carry out a reaction cycle involving brominating polyethylene reducing it to the original hydrocarbon and then brominating the polymer once again. Bromination of both LDPE and HDPE gives rise to large increases in adhesion. Reduction of the brominated polymer by TBTH causes the adhesion to fall to the level of the untreated polymer. The reduced surface can then be rebrominated and high adhesion values are again obtained.

Tables 2 and 3 show the adhesion measurements and XPS data of this cycle of reactions with both LDPE and HDPE. It has been possible to achieve low levels of oxygen (<0.7%) in these samples, which is comparable with the oxygen level in untreated polyethylene. This shows therefore, that the change in adhesion levels corresponds essentially to the change in bromine content. The results in Tables 2 and 3 strongly indicate that the basic reason for the poor adhesion of polyethylene is a lack of functional groups and not weak boundary layers.

3.3 Dehydrohalogenation of Brominated Polyethylene

A cycle similar to that obtained by reducing the brominated polymer may be carried out by dehydrohalogenating the brominated polymer, and then rebrominating the resulting double bonds in the dark. The reaction sequence is shown in schemes 1 and 2. The numbers in brackets refer to the joint strengths in MN m^{-2}; further data are given in Tables 4 and 5.

Table 2 - Bromination - reduction cycle with LDPE

	Joint Strength/ MN m^{-2} ± s dev	Failure	Equilibrium Contact Angle	XPS Data (atomic ratio) C:Br
Untreated LDPE	1.4 ± 10%	I	96°	∞
Brominated LDPE (2 hour irradiation)	12.7 ± 13%	I/M	76°	9:1
TBTH reduction	1.8 ± 7%	I	94°	∞
Rebrominated (2 hour irradiation)	10.5 ± 11%	I	78°	12:1
M = material failure				

Table 3 - Bromination - reduction cycle with HDPE

	Joint Strength/ MN m^{-2} ± s dev	Failure	Equilibrium Contact Angle	XPS Data (atomic ratio) C:Br
Untreated HDPE	0.7 ± 9%	I	96°	∞
Brominated HDPE (3½ hour irradiation)	11.5 ± 14%	M	74°	17:1
TBTH reduction	0.76 ± 13%	I	94°	∞
Rebrominated (3½ hour irradiation)	10.0 ± 9%	M	72°	26:1

Table 4 - Bromination - dehydrobromination - rebromination cycle for LDPE

	Joint Strength/ MN m^{-2} ± s dev	Failure	Equilibrium Contact Angle	XPS Data (atomic ratio) C:Br
Untreated LDPE	1.4 ± 10%	I	96°	∞
Brominated LDPE (1 hour irradiation)	6.7 ± 7%	I	73°	22:1
Unsaturated LDPE	3.7 ± 11%	I	65°	715:1
Unsaturated LDPE washed CHCl$_3$	3.6 ± 11%	I	65°	719:1
Rebrominated unsaturated LDPE (in dark)	6.3 ± 13%	I	58°	29:1

Table 5 - Bromination - dehydrobromination - rebromination cycle for HDPE

	Joint Strength/ MN m^{-2} ± s dev	Failure	Equilibrium Contact Angle	XPS Data (atomic ratio) C:Br
Untreated HDPE	0.7 ± 9%	I	96°	∞
Brominated HDPE 3½ hour irradiation)	11.2 ± 12%	I	75°	19:1
Unsaturated HDPE	2.9 ± 13%	I	67°	200:1
Rebrominated unsaturated HDPE	15.9 ± 6%	M	55°	12:1

Scheme 1 - Dehydrobromination Rebromination of LDPE

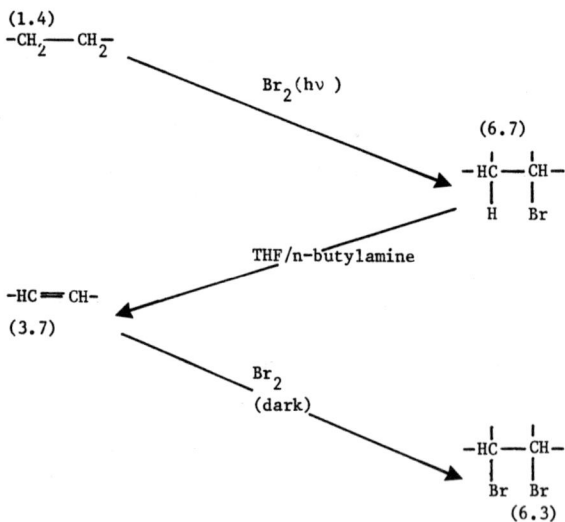

Scheme 2 - Dehydrobromination   Rebromination of HDPE

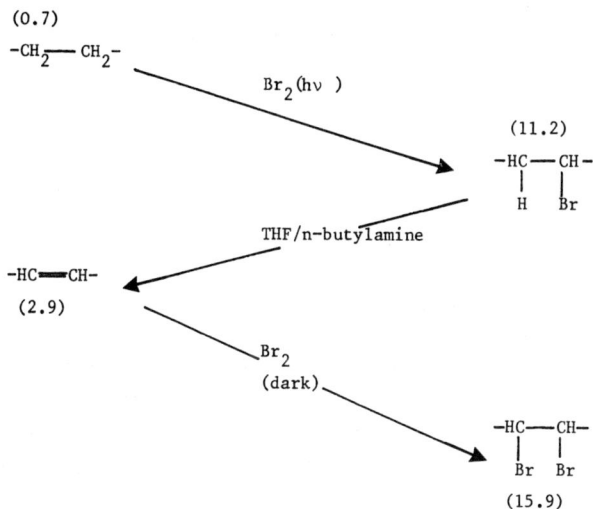

(0.7)
$-CH_2-CH_2-$

$Br_2(h\nu)$

(11.2)
$-HC-CH-$
         | |
         H   Br

THF/n-butylamine

$-HC=CH-$
(2.9)

Br_2
(dark)

$-HC-CH-$
  | |
  Br   Br
(15.9)

After the initial bromination the polymers are treated in THF/
n-butylamine for one week. The films are washed in pure THF, followed by
dilute hydrochloric acid and water.

This treatment introduces unsaturation into both LDPE and HDPE
surfaces, the adhesion being approximately three times that of the original
untreated polymers. The double bonds may be broken by exposing the films
to bromine vapour, in the dark. This should introduce twice the amount of
bromine that was originally present and hence give a higher joint strength.
This is indeed the case with HDPE, but it was necessary to remove low
molecular weight material from LDPE before an improvement in adhesion was
observed. However, bromination of the double bonds in LDPE gives only the
same level of bromine as the original brominated polymer. This is reflected
in the corresponding joint strength. The reasons for this are not understood
and work will be carried out to investigate this further.

3.4 Reactions of Unsaturated Polyethylene Surfaces

Unsaturated polyethylene was reacted with several different reagents
in order to introduce specific functional groups into the polymer surface.
The resultant films were then derivatised with each of three reagents that
should label a specific functional group[3,7].

The derivatisation reagents react with specific functional groups in the following manner.

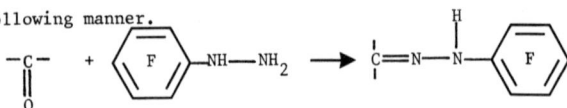

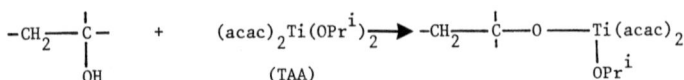

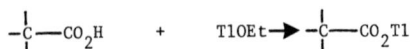

$$-\underset{\underset{O}{\parallel}}{C}- \quad + \quad \langle\bigcirc\rangle\text{-F} \quad -NH-NH_2 \quad \longrightarrow \quad \underset{}{C}=N-\overset{H}{\underset{}{N}}-\langle\bigcirc\rangle\text{-F}$$

$$-CH_2-\underset{\underset{OH}{|}}{C}- \quad + \quad (acac)_2 Ti(OPr^i)_2 \longrightarrow -CH_2-\underset{}{C}-O-\underset{\underset{OPr^i}{|}}{Ti}(acac)_2$$
$$(TAA)$$

$$-\underset{}{C}-CO_2H \quad + \quad TlOEt \longrightarrow -\underset{}{C}-CO_2Tl$$

Although TlOEt reacts very well with carboxylic acid groups it has been shown in the present work that it also reacts with hydroxyls. This was confirmed by hydrolysing a co-polymer that contained acetate groups. This produced free -OH groups on the surface, and the polymer was treated with TlOEt. XPS showed that large amounts of Tl were present on the surface, and so this reagent is not sufficiently specific for the purposes of this work.

a) Hydroboration - Oxidation

This reaction was carried out in order to introduce a single hydroxyl group into the surface of polyethylene for each double bond.

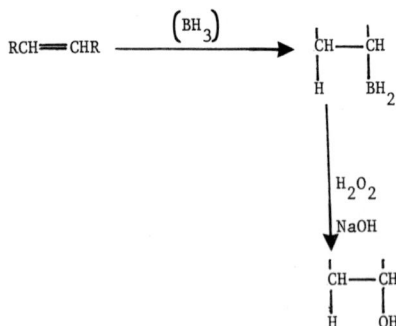

$$RCH=CHR \quad \xrightarrow{\left(BH_3\right)} \quad \underset{\underset{H}{|}}{CH}-\underset{\underset{BH_2}{|}}{CH}$$

$$\downarrow {\substack{H_2O_2 \\ NaOH}}$$

$$\underset{\underset{H}{|}}{CH}-\underset{\underset{OH}{|}}{CH}$$

This reaction appears to introduce a large number of -OH groups into the surface of LDPE, but not HDPE. Considerable amounts of Ti can be seen by XPS after derivatisation with TAA. Very small amounts of F can be seen after derivatisation with PFPH, indicating few carboxyl groups. It is not

Table 6 - Hydroboration - oxidation and derivatisation reactions

	Joint Strength/ MN m^{-2} ± s dev	Type of Failure	Equilibrium Contact Angle	XPS data/Atomic Ratios				
				C:Br	C:O	C:F	C:Ti	C:Tl
Original brominated LDPE	6.0 ± 12%	I	79°	22:1	40:1	-	-	-
Unsaturated LDPE	3.9 ± 7%	I	68°	369:1	38:1	-	-	-
Reaction with BH$_3$	4.6 ± 15%	I	62°	-	17:1	-	-	-
BH$_3$ + TAA				-	21:1	-	66:1	224:1
BH$_3$ + PFPH				-	42:1	220:1	-	-
BH$_3$ + TlOEt				-	19:1	-	-	24:1
Untreated LDPE reacted with BH$_3$				-	96:1	-	-	-

certain whether this reaction introduces a significant number of carboxylic acid groups, as TlOEt has been shown to react with both hydroxyls and carboxylic acids. Table 6 shows the XPS data and adhesion measurements.

b) Potassium Permanganate – Oxidation

This reaction was carried out in order to introduce vicinal hydroxyl groups into the surface of polyethylene.

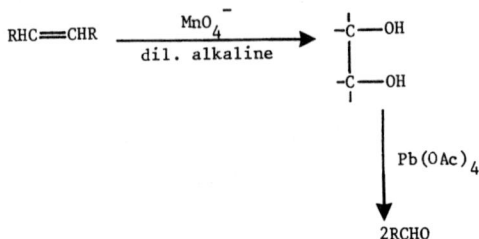

$$RHC{=\!\!=}CHR \xrightarrow[\text{dil. alkaline}]{MnO_4^-} \begin{array}{l} -\overset{|}{C}-OH \\ -\overset{|}{C}-OH \end{array}$$

$$\downarrow Pb(OAc)_4$$

$$2RCHO$$

As in the previous reaction, large quantities of oxygen were seen by XPS after treating LDPE in potassium permanganate, but none was found on the surface of HDPE. The oxidised LDPE was derivatised with TAA but no Ti was detected by XPS. It is most probable that this is due to the TAA not reacting with a pair of hydroxyl groups in the same way that it reacts with a single group. The adhesion of this polymer to an epoxide adhesive was comparable to that of the film containing single hydroxyl groups (6.2 MN m^{-2}) and the equilibrium contact angle of water on the surface was 58°.

It is hoped in the near future to cleave the 'diol' with lead tetracetate to produce an aldehyde. This could be unambiguously identified by derivatising with PFPH. It will also be very interesting to be able to compare the adhesion properties of films containing ketonic carbonyl groups with those containing aldehydic carbonyl groups.

c) Permanganate/Periodate – Oxidation

The purpose of this reaction is to introduce carboxylic acid groups into polyethylene surfaces.

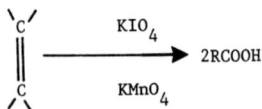

$$\begin{array}{c} \diagdown C \diagup \\ \| \\ C \\ \diagup \diagdown \end{array} \xrightarrow[KMnO_4]{KIO_4} 2RCOOH$$

The double bond is oxidised to the diol by permanganate which is then cleaved to give aldehydes and ketones. Any aldehydes are then further

Table 7 - Derivatisation of Carboxylic Acid Groups on LDPE

	Joint Strength/ MN m^{-2} ± s dev	Type of Failure	Equilibrium Contact Angle	XPS Data/Atomic Ratios				
				C:Br	C:O	C:F	C:Ti	C:Tl
Original brominated LDPE	6.3 ± 7%	I	80°	27:1	86:1	-	-	-
Unsaturated LDPE	4.2 ± 11%	I	70°	-	66:1	-	-	-
Unsaturated LDPE treated with $KMnO_4/KIO_4$	11.2 ± 13%	I/C	64°	-	6:1	-	-	-
$KMnO_4/KIO_4$ + TAA				-	10:1	-	196:1	210:1
$KMnO_4/KIO_4$ + PFPH				-	39:1	189:1	-	-
$KMnO_4/KIO_4$ + TlOEt				-	9:1	-	-	28:1
Untreated LDPE + $KMnO_4/KIO_4$	2.2 ± 10%	I		-	42:1	-	-	-

Table 8 - Derivatisation of Carboxylic Acid Groups on HDPE

	Joint Strength/ MN m^{-2} ± s dev	Type of Failure	Equilibrium Contact Angle	XPS Data/Atomic Ratios				
				C:Br	C:O	C:F	C:Ti	C:Tl
Original brominated HDPE	6.6 ± 12%	I		24:1	163:1	-	-	-
Unsaturated HDPE	2.4 ± 6%	I		162:1	160:1	-	-	-
Unsaturated HDPE treated with KMnO$_4$/KIO$_4$	7.1 ± 6%	I		-	28:1	-	-	-
KMnO$_4$/KIO$_4$ + TAA				-	22:1	-	254:1	-
KMnO$_4$/KIO$_4$ + PFPH				-	25:1	477:1	-	-
KMnO$_4$/KIO$_4$ + TlOEt				-	26:1	-	-	49:1

oxidised to carboxylic acids. This appears to have been successful with both LDPE and HDPE. As can be seen from Tables 7 and 8 large quantities of T1 have been detected but very small amounts of F and Ti are present. This indicates that the majority of the oxygen atoms on the surface are present as carboxylic acid groups and there is a very small quantity of hydroxyl and carbonyl groups.

The adhesion of both the oxidised LDPE and HDPE is very high compared to the original untreated polymer. This therefore indicates that carboxylic acid groups cause large increases in the adhesion of the polymers to an epoxide adhesive. The equilibrium contact angle of water on the surfaces are 64° and 58° on LDPE and HDPE respectively.

3.5 Silver Tetrafluoroborate

Silver tetrafluoroborate has been successfully used in introducing ketonic carbonyl groups into the surface of both LDPE and HDPE.

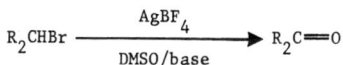

$$R_2CHBr \xrightarrow[\text{DMSO/base}]{\text{AgBF}_4} R_2C{=}0$$

The XP spectrum of LDPE shows that there is still a small amount of residual bromine, but there is a considerable amount of oxygen on the surface. Derivatisation with PFPH caused the oxygen signal to be reduced and a fluorine signal to be observed. This signal corresponded to five fluorines per oxygen atom lost.

The adhesion of both LDPE and HDPE containing ketonic carbonyls, to an epoxide adhesive was very high compared to the untreated polymers. It is therefore likely that keto groups introduced by various pretreatments play an important role in the resulting enhanced adhesion.

Derivatisation of HDPE with the three derivatisation reagents showed there to be a majority of keto groups on the surface. Unfortunately, this was only carried out qualitatively but will be carried out quantitatively in the near future.

4 SUMMARY

The results are summarised in scheme 3 and Table 9.

Table 9 - Summary of Joint Strengths of Treated LDPE and HDPE

	Joint Strengths $(MN\ m^{-2})$	
	LDPE	HDPE
Untreated	1.4	0.7
Brominated	6.7	5.4
Unsaturated	3.7	2.9
$-\overset{\mid}{C}-OH$ / $-\overset{\mid}{C}-H$	4.6	-
$-\overset{\mid}{C}-OH$ / $-\overset{\mid}{C}-OH$	4.9	-
$-COOH$	11.3	7.1
$\overset{\mid}{C}=O$	15.7	9.3

Scheme 3

$$-CH_2-CH_2-$$

TBTH $Br_2(h\nu)$

$$-HC-CH-$$
$$\quad | \quad |$$
$$\quad H \quad Br$$

$AgBF_4$ n-butylamine/THF

DMSO

$$-HC-C-$$
$$\quad | \quad \|$$
$$\quad H \quad O$$

$$-HC=CH-$$

dilalkaline $KMnO_4$ "BH_3"/H_2O_2 $KMnO_4/KIO_4$

$$HC-OH$$
$$|$$
$$HC-OH$$

$$-HC-CH-$$
$$\quad | \quad |$$
$$\quad H \quad OH$$

$$2-COOH$$

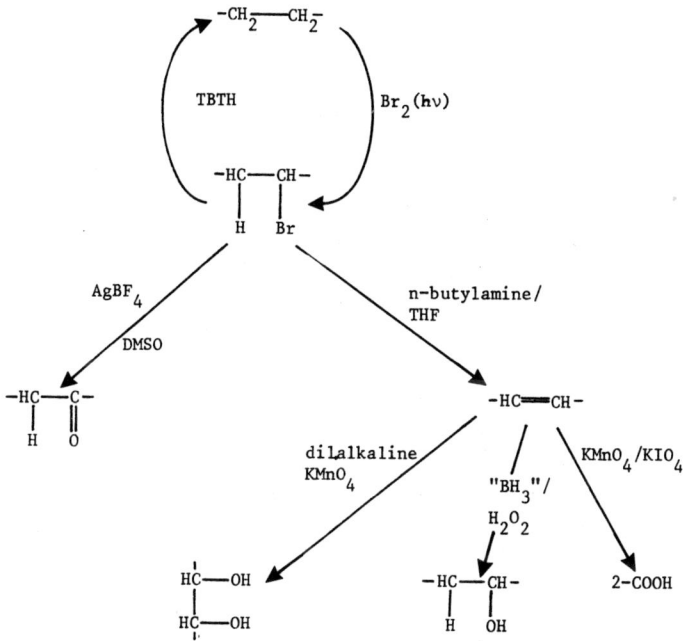

5   CONCLUSIONS

   a)   The bromination reduction cycles provide strong evidence to show
        that the basic reason for the poor adhesion with polyethylene
        is a lack of functional groups and not weak boundary layers. This
        does not mean that weak boundary layers are never important with
        polyolefins.

   b)   At least for the epoxide adhesive under examination, bromo, keto,
        and carboxylic acid groups are considerably more effective in
        promoting adhesion than alkene or hydroxyl groups.

REFERENCES

1. Briggs, D. Chapter 9 in "Surface analysis and surface pretreatments of plastics and metals", Brewis, D M. (ed.) Applied Science Publishers, 1982.

2. Brewis, D M and Briggs, D. "Adhesion to polyethylene and polypropylene", Polymer, Vol. 22, 7, 1981.

3. Briggs, D and Kendall, C R. "Derivatisation of discharge treated LDPE: An extension of XPS analysis and a probe of specific interactions in adhesion", Int. J. Adhes. and Adhes., Vol. 2, No. 1, 13, 1982.

4. Perrin, D D, Armarego, W L F and Perrin, D R. "Purification of laboratory chemicals", Pergamon Press, 1980.

5. Sawyer, D T and Roberts, J L. "Experimental electrochemistry for chemists", Wiley - Interscience, 1974.

6. Vogel, "Textbook of practical organic chemistry", Longman, 1978.

7. Baitch, C D and Wendt, R C. "Photon, electron and ion probes of polymer structure and properties", Dwight, Ralish and Thomas (eds.), 221, Washington American Chemical Society, 1981.

8. Briggs, D, Zichy, V J I, Brewis, D M, Comyn, J, Dahm, R H, Green, M A and Konieczko, M B. "X-Ray photoelectron spectroscopy studies of polymer surfaces", Sur. and Int. Anal., Vol. 2, No. 3, 107, 1980.

Chapter 7

CHARACTERISTICS OF TOUGHENED ACRYLATE ADHESIVES USED WITH ALUMINIUM ALLOY
ADHERENDS

C. BURROWS*, N. SAMMES[+], G. STABLES*, M.H. STONE[+] and C. TEMPEST*

*Lucas Aerospace Ltd, Burnley, Lancs
+Materials and Structures Dept., Royal Aircraft Establishment, Farnborough,
Hants

SYNOPSIS

      The two-part, toughened acrylate adhesives have a potential role in rapid
field repairs of aircraft components. The properties and application techniques
of the acrylates differ from those of epoxy adhesives, and this evaluation
illustrates the distinctive characteristics of acrylate adhesives when used to
bond aluminium alloys.

      The effect of various surface contaminants and the use of simple metal
surface pre-treatments were examined. Acceptable joint strengths were obtained
even with adherends deliberately contaminated with aircraft fluids (including
silicone oil), then pre-treated merely by dry wiping and abrading. Some
acrylates show acceptable strength at 80°C but strengths at 120°C were low. At
-55°C lap shear strengths were acceptable but peel strengths were very low,
indicating brittleness at this temperature. Cure rates varied appreciably with
the composition of aluminium alloy adherends.

      Variability of strength tended to be higher than for epoxy adhesives
despite control of obvious variables such as joint open time, glue line
thickness, mode of application of adhesive, cure temperature and time.
Furthermore, it did not appear to be an effect of oxygen inhibition nor a
function of atmospheric humidity during lay-up.

1   INTRODUCTION

      There is a need for adhesives suitable for rapid, temporary repairs to
aircraft structures in difficult field conditions, carried out by personnel

without special skills in adhesive bonding.   Such adhesives must therefore
be tolerant of imperfections in techniques, such as errors in mix ratio,
incomplete mixing and poorly prepared surfaces.

Cold-setting epoxy adhesives have shortcomings for this kind of job.
They need reasonably complete and accurate mixing, they give very weak bonds to
oily surfaces, and if the cure is fast enough for the purpose (2 to 6h at 20°C)
they have inconveniently short work lives.   Also, their heat distortion
temperature is generally low even when post-cured above ambient temperature.

In contrast, acrylate adhesives offer a number of potential advantages.

(a)   They cure by a free-radical mechanism and so the properties are
less dependent on accurate and complete mixing.

(b)   Many examples of the type give acceptable bond strengths with oily
surfaces.

(c)   A reasonably long work life (eg 10 to 20 min) is followed by a rapid
cure;   the partly anaerobic character of these adhesives has been deliberately
accentuated in some examples in order to enhance this favourable ratio of cure
time to work life.   Also, a variety of curing systems is available, offering
a very wide range of work lives and cure times for different needs.

(d)   The acrylates tend to have higher strength than cold-setting
epoxies at moderately elevated temperatures (up to about 80 to 100°C).

Several useful general descriptions of acrylate adhesives have been
published by manufacturers[1-5], but details of composition can only be inferred
from the patent literature much of which was reviewed by Prane[6].   Essentially
the acrylate adhesives comprise solutions of toughening rubbers and acrylate
polymers in acrylate monomers (mainly methyl methacrylate), usually with a
minor amount of acrylic or methacrylic acid which increases bond strength and
cure speed.   Recent patents indicate a concern to improve heat resistance by
changing the toughening rubber[7-9], and to increase adhesion to poorly-prepared
and contaminated metals by including specific adhesive promoters[10].   The curing
system typically consists of:

(a)   a free-radical generator, usually one of the organic peroxides
commonly used for vinyl polymerisation;

(b)   an "initiator", often a tertiary amine;

and, in some cases

    (c)   an "accelerator", such as a transition metal salt.

The components of the cure system are divided between the resin base and a "catalyst" solution that are separately storage stable.   The resin and catalyst may be mixed before application of the adhesive; or in some formulations they are applied separately to the parts to be bonded and then mix adequately when the parts are assembled.

    Cure results in polymerisation of the acrylate monomers and cross-linking via reactive groups on the rubber, which is precipitated as a dispersed phase in an acrylate matrix; cross-linking may also occur via di-acrylate esters in some formulations.

    Thus the acrylate adhesives differ substantially from their epoxide counterparts, and their distinctive characteristics must be understood in order to use them effectively.   In the work reported here the adherends were always aluminium alloys.   An important aim of the study was to determine strengths obtained with contaminated adherends that were given the minimum of surface preparation, ie abrasion with or without solvent cleaning.   Not all the adhesives proceeded through the full evaluation: some were withdrawn from sale while others proved unsuitable for our purpose.

    Four main aspects were studied:

    (a)   variability of joint strengths;

    (b)   effect of contaminated surfaces;

    (c)   effect of test temperature;

and   (d)   effect of adherend type.

2   EXPERIMENTAL METHODS

    For clarity each results table includes a note of the alloys and adhesives, surface treatments and cure conditions used.   Fuller descriptions are given below.

2.1 Materials

2.1.1 Alloys: Most of the work was carried out with clad aluminium alloy to BS L73 (or the later version BS L165).   However, a comparison of joint strengths and cure rates was made between this alloy and alloys to BS L157 (unclad version of L73) and 7075-T6 clad and unclad.

2.1.2 <u>Adhesives</u>: The adhesives fall into two obvious groups, the "premix" type (eg H, I, O1, P2) where the "catalyst" was mixed with the adhesive before the mixture was applied to the adherends; and the "separate-coat" type (eg C, D, M, N, O, P) where the resin base and catalyst were intended to be applied separately to opposing adherends. Some adhesives could be used in either way, in some cases with different catalysts for the different methods of application. The separate-coat catalysts were of two types, an oily, non-volatile solution of a tertiary amine or a solution of peroxide in volatile solvent that dried to a tack-free film.

2.2 <u>Surface Pretreatments of Adherends</u>

2.2.1 <u>Chromic/sulphuric acid pickle</u>: Initially it was desired to measure the inherent strengths of the adhesives, free of any effects of reduced adhesion caused by minimal surface preparation. For this purpose chromic/ sulphuric acid pickle to Ministry of Defence DEF-STAN 03-2/1 Method 0 was used, preceded by vapour degreasing and aqueous alkaline cleaning.

2.2.2 <u>Solvent cleaning and abrasion</u>: Some variations in technique occurred as the programme evolved and are noted in the results tables. In earlier work the swab was an open weave cloth, the solvent 1,1,1-trichloroethane, and abrasion was by hand with 240 grade waterproof paper.

In a later, more thorough examination of the effect of this form of treatment materials were chosen that were more likely to be used in practice. The process comprised swabbing with tissues soaked in paint thinners (Ministry of Supply Specification DTD 840, 90% mixed isomers of xylene, 10% n-butanol by volume), abrading with a 150 grade "flap wheel" (fan grinder), and then swabbing again. By "swabbing" we mean that the surface was first liberally flooded with solvent which was allowed to act for up to about a minute, and then a dry tissue was used to wipe off the film of solvent plus dissolved contaminants. Disposable polyethylene gloves were worn. Fresh solvent, tissues and gloves were used for the post-abrasion swabbing to avoid carrying over contamination. When different alloys were being compared separate flap-wheels were reserved for each alloy.

2.2.3 <u>Dry wiping and abrasion</u>: This comprised simply wiping with dry tissue, abrading with a flap-wheel, and then wiping again with a fresh dry tissue.

2.2.4 <u>Deliberate contamination before pre-treatment</u>: In some of the earlier work single oils were applied to pickled surfaces before swabbing and abrasion. In

a later, more thorough study of the effect of prior contamination, mixtures
were made to give 4 classes of fluids or pastes likely to be found on aircraft:
oils, greases, corrosion preventives and de-icing fluids (described more
fully in section 3.2.3).   Equal parts of these 4 mixtures were also combined
to form a "grand contaminant mixture" for some of the work.   These mixtures
were applied to vapour degreased metal prior to the cleaning and abrasion
described above in order to simulate field conditions.   To avoid cross
contamination separate flap-wheels were reserved for use with each of the
contaminant groups.

2.3 Bonding

Joints were bonded within 4h of the completion of pretreatment.   About
2% by weight of glass ballotini of size 70-90 μm were added to the adhesive for
glue line thickness control.   The adhesive was normally applied as a "bead",
ie a thick, narrow band along the adherend, in order to minimise monomer loss
by evaporation; variations from this technique are noted in the results
tables.   With the separate-coat adhesives the catalyst was brushed on to
give a thin film.   In the later work with pre-mix types P2 and 01 the mixing
was done with a plastic spatula to eliminate any possible effect of transition
metal ions affecting the cure.   The time between spreading the adhesive and
closing the joint was controlled and was usually either 1, 2 or 15 min.   After
joint assembly pressure was immediately applied by weights to give 15-35 kPa.

Cure was normally at ambient temperature which in later work was
controlled at $20 \pm 2\,°C$; a limited study of post-cure at 80°C was also made.
Cure times ranged from 2h to 12 days during the various studies, but were
always controlled to eliminate this as a variable.

2.4 Joint Types and Tests

After fracture all joints were visually examined to assess the proportions
of adhesion and cohesive failure.

2.4.1 Lap shear strength: Conventional single overlap joints 25.0 or 25.4 mm
wide and 11.0 or 12.5 mm overlap were made with 2 mm thick alloy.   Care was
taken to minimise any post-cure that might occur from heating during cutting
of joints from the bonded panels; in some cases the panels were pre-slotted to
avoid cutting through the bonded area.   The joints were tested at a crosshead
speed of 1 mm/min (RAE work) or 4.4 kN/min (Lucas work).   In the most recent
work with contaminated and abraded adherends the tests at elevated temperatures
were started as soon as the joints reached the required temperature, rather

than allowing the customary 15 min wait before applying load. This was done to minimise post-cure of the adhesive at the test temperature and to give a more realistic simulation of the conditions experienced when an aircraft accelerates to supersonic speeds. Typically 5-7 replicates were tested.

2.4.2 Peel strength: 90° peel strength was determined essentially as described in Ministry of Aviation Specification DTD 5577, except that the alloys of interest were used rather than the specified BS L61, and the joints were made separately rather than cut from panels. Mean peel strengths were calculated from readings at intervals along the peel trace, and the results shown were means of usually three joints.

2.4.3 Wedge cleavage test for toughness and durability: The joints were essentially as described in specification ASTM D3762, except that in most of the work they were made as separate joints rather than as a panel in order to reduce the risk of large voids. Three to six replicates were tested. After driving in the wedge the initial crack length was measured within 1h. The joints were then exposed for 45 ± 2h at 50°C/95% rh and crack length re-measured; finally they were split open for assessment of failure mode.

Results are reported as fracture energy before and after exposure. Fracture energy was calculated from the approximate equation,

$$G = \frac{3Ed^2h^3}{16a^4}$$

Values of G must be regarded as only semi-quantitative mainly because some yielding of the adherends occurred. Nevertheless, G is considered to be more informative than crack length or crack extension in revealing differences between adhesive/surface treatment combinations.

2.5 Monomer Loss by Evaporation

Adhesive was applied to alloy plates placed in an airstream of 0.5 m/s to simulate a working environment (typical fume cupboard air speed); weight loss was determined after 15 min exposure, and with one exception the adhesives were uncatalysed. The adhesives were normally applied as a bead, but a few measurements were also made with uniform films 50 x 50 mm in area and 0.675 mm thick. Data are means of three replicates.

3 RESULTS AND DISCUSSION

3.1 Joint Manufacture Techniques and Variability of Joint Strength

Variability of joint strength with the acrylate adhesives tended to be higher than in the authors' experience with epoxy adhesives despite control of obvious variables such as cure time and temperature, and exclusion of voidy joints from the results. Also, care was taken not to disturb assembled joints during the early stages of cure (a well-established cause of reduced strength), and there was no consistent effect of glue line thickness over the range observed (approximately 0.05-0.4 mm). Other possible causes of variable strengths were therefore studied.

3.1.1 Monomer loss during open assembly time: It is well known that appreciable monomer loss can occur from applied adhesive prior to joint assembly, and tab. 1 gives the results obtained.

Table 1:

Monomer loss from adhesive applied to a plate, over 15 minutes at an air speed of 0.5 m/s

Adhesive	A	A[1]	B	C	D	E	F	H	I	J	K	M	P2[2]
Weight Loss (%)	13	29	12	16	18	19	0.5	2.5	9.4	2.6	1.9	1.4	7.9

Note: (1) from a film; (2) catalysed adhesive

The wide range of volatility for this group of materials suggests that there should also be a wide range of strength reductions with extended open time. The higher losses shown in tab. 1 are a substantial, even major, proportion of the total free monomers present, and high losses would give a weaker, more rubbery adhesive with poor wetting characteristics and an altered phase structure.

The effect on lap shear strength of 15 minutes open time compared to minimum open time (about 0.5 min) is shown in tab. 2.

Adhesives H, I and M, with low to medium monomer loss (tab. 1), showed no significant effect of open time, whereas the high loss adhesives C and D gave weaker joints after 15 minutes open time. The first set with adhesive C (a separate coat type) gave over 50% strength loss for 15 minutes open time, and differences in colour of the fracture surfaces indicated

that the initiator had failed to mix well with the adhesive.   A second
panel did not show this effect.   Other data for adhesive C also show an
appreciable effect of 15 minutes open time[11].

Table 3 shows the effect of open time on peel strength under the same
conditions as for the results in tab. 2.

Table 2:

Effect on lap shear strength of open time
in an air stream of 0.5 m/s

Adhesive	Minimum open time		15 minutes open time	
	Strength (MPa)	Standard deviation (MPa)	Strength (MPa)	Standard deviation (MPa)
C	26.4	0.40	11.1	0.73
C (repeat)	-	-	24.3	1.26
D	33.2	1.42	28.3	0.66
H	32.6	0.69	31.5	1.30
I	29.5	1.30	28.6	1.86
M	33.9	3.41	32.8	1.92

Note: Etched L73 adherends, 1-3 day cure.

Table 3:

Effect on peel load of open time in an air stream
of 0.5 m/s

Adhesive	Minimum open time		15 minutes open time	
	Peel load (N/mm)	Standard deviation (N/mm)	Peel load (N/mm)	Standard deviation (N/mm)
C	8.0	0.87	6.3	2.86(1)
D	1.9	0.89(1)	1.8	0.62
H	7.8	0.88	6.3	0.56
I	5.4	0.40	5.4	0.79(1)
M	7.5	1.18	9.9	0.93

Note: etched L73 adherends, 1-3 day cure;   (1) 2 joints only

There were no marked effects of open time on strength although the
increase for Adhesive M is consistent with a higher proportion of rubbery
component giving a tougher adhesive.   However, the effect of monomer loss on
wetting and flow were indicated by an increased tendency for large voids to be
formed upon assembly after 15 minutes open time.

Thus, although the effect of exposure before assembly is perhaps not as
large as might have been expected from the monomer loss results (tab. 1), an
extended open time is clearly undesirable.   In all the other work reported
here it was fixed at one or two minutes.

3.1.2   Effect on peel strength of applying the catalyst to the peel strip: In
the 90° peel test the stress is concentrated at the surface of the thin peel
strip.   It therefore seemed likely that the peeling force could depend on
which component of a separate–coat adhesive (the adhesive base or the catalyst)
was applied to the peel strip.   Adhesives C and M were tested for this effect,
with the results shown in tab. 4.

Table 4:

Effect on peel strength of applying the catalyst
to the thin peel strip

Adhesive	Adhesive on peel strip		Catalyst on peel strip	
	Peel load (N/mm)	Standard deviation (N/mm)	Peel load (N/mm)	Standard deviation (N/mm)
C	8.0[1]	0.41	7.2[1]	0.43
M	9.0[2]	1.28	Nil	–

Note: etched L73 adherends, 1-3 day cure;   (1) 3 replicates, (2) 12 replicates

The complete contrast between Adhesives C and M is probably due to the
different forms of the catalyst component.   For adhesive C the catalyst
component is a thin oil containing the tertiary amine part of the cure system;
for Adhesive M the catalyst contains the peroxide part of the cure system
and after application is present as a thin dry film in which crystals of
peroxide are visible.   Presumably this film-forming resin does not completely
dissolve and disperse uniformly throughout the adhesive thickness, whereas the
oily film of Adhesive C is much more likely to do so.   For all the other work
reported here the adhesive base was applied to the thin peel strip.

3.1.3 Effect of adhesive/catalyst ratio with separate-coat adhesive types:
With the separate-coat type of adhesive the concentration of catalyst in the
completed joint will vary with both the final adhesive layer thickness and with
the relative amounts of adhesive and catalyst applied. It is well known
that strength decreases as the glue line thickness increases, because the
catalyst concentration is lower and the reacting species do not completely
diffuse across the thick adhesive layer. However, as noted earlier, no clear-
cut and consistent effect of glue line thickness was observed in this work, in
which thicknesses were less than 0.5 mm. The catalyst concentration was
therefore varied instead by deliberately applying thinner or thicker coats
than usual, to simulate the kind of differences that might be found between
different operators. The results are shown in fig.1 for Adhesive C and fig.2
for Adhesive M. The catalyst concentration for Adhesive C was derived from the
weights of adhesive and catalyst applied to opposing adherends, and was not
the true catalyst concentration in the cured joint. For Adhesive M the
catalyst weight is plotted, because no excess of this dry film catalyst is
expelled from the joint as pressure is applied and the concentration is
therefore approximately proportional to the weight applied. For both adhesives
there is no real trend of either lap shear or peel strength with catalyst
amount, and normal variations in weight of catalyst applied are unlikely to
be a substantial cause of variability (all joints in this work were made by
the same operator).

3.1.4 Effect on peel load of the presence of oxygen during mixing of adhesive
and lay-up of joints: The acrylate adhesives are partly anaerobic, as shown by
incomplete cure of adhesive spew around a joint. Oxygen incorporated during
mixing and lay-up might tend to inhibit cure, giving variable strengths
depending on how the adhesive was mixed and applied. Joints made in air and
nitrogen were therefore compared (N_2 contained 0.4-2.5% O_2). To exaggerate
any effect the adhesive mixing in air was done in such a way as to incorporate
as much air as possible. The adhesive was also applied in different ways:

    (a) as a bead down the centre of the peel strip;
    (b) thickly all over the peel strip;
or, (c) thickly all over the back plate.

These techniques gave varying degrees of oxygen exposure to the adhesive in
contact with the peeling interface, as well as revealing any effects of
"skinning" of the adhesive caused by monomer evaporation. The results are
shown in tab. 5 for the pre-mix type adhesives 01 and P2.

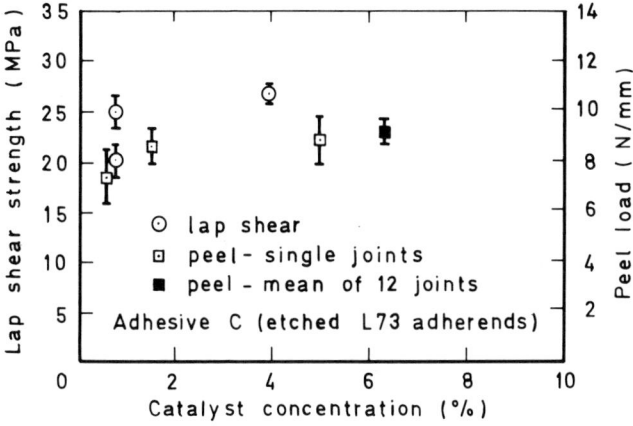

Fig.1 Effect of catalyst concentration on lap
shear and peel strengths for adhesive C

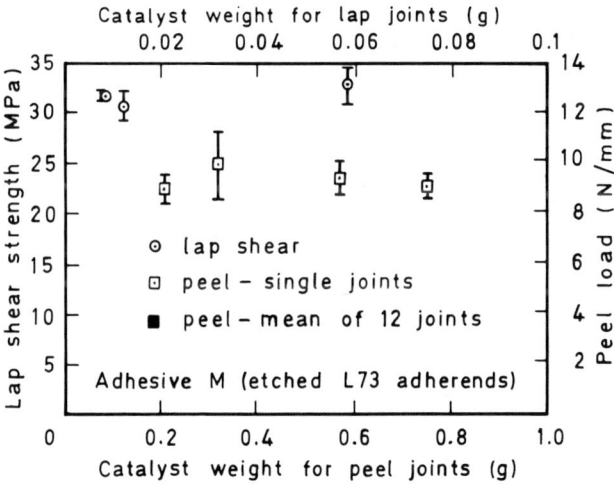

Fig.2 Effect of catalyst weight on lap shear
and peel strengths for adhesive M

Table 5:

Effect on peel load of atmosphere during lay-up and mode of
adhesive application

Adhesive	Atmosphere	Mode of application	Peel load (N/mm)	
			Load	Standard deviation
01	air	bead on peel spread on peel spread on backing	5.7 6.0 5.2	0.12 0.63 0.21
	nitrogen	bead on peel spread on peel spread on backing	5.2 5.7 4.7	0.16 0.56 0.30
P2	air	bead on peel spread on peel spread on backing	9.5 9.8 9.5	0.53 0.44 0.12
_	nitrogen	bead on peel spread on peel spread on backing	9.5 9.7 6.9	0.36 0.13 0.59

Note: etched L73 adherends, 3h cure

It is clear that varying degrees of oxygenation are unlikely to cause
significant variability with these adhesives.   Also, provided the adhesive
was applied to the peel strip the exact manner was unimportant: however when
spread on the backing plate there was a tendency for lower strengths, most
noticeably for P2 in nitrogen.

3.1.5 Effect on peel load of relative humidity during mixing and lay-up:
It was thought possible that atmospheric humidity was affecting either the cure
or the surface.   Any surface effect might have differed according to the pre-
treatment, so both etch and abrade/swab treatments were used in a study of the
effect of relative humidity.   The pretreated adherends were placed in the
controlled humidity for 1 hour before bonding, and the adhesive was mixed and
the joints laid up in the same atmosphere.   The results for adhesives 01 and P
are shown in tab. 6.

127

Table 6:

Effect on peel load of relative humidity during bonding

Adhesive	Pre-treatment	Relative humidity (%)	Peel load (N/mm)	
			Load	Standard deviation
01	abrade, swab	77	6.1	0.13
		54	5.5	0.37
		45	5.9	0.59
		4	6.4	0.20
	etch	77	5.2	0.50
		54	5.2	0.13
		45	5.5	0.19
		4	6.4	0.34
P2	abrade, swab	77	2.3	0.16
		49	2.4	0.11
		48	2.4	0.19
		3.5	3.0	0.12
	etch	77	9.8	0.41
		49	9.5	0.46
		48	9.6	0.55
		3.5	9.6	0.44

Note: L73 adherends, 3h cure

It is evident that there are no major effects of relative humidity. For Adhesive 01 with etched adherends and Adhesive P2 with abraded adherends there is a possible tendency to higher strength in dry air; but ordinary variations of relative humidity in a laboratory atmosphere are unlikely to be a significant cause of variability with these adhesives.

3.1.6 Causes of variability: Although several possible causes of variability have been identified none could have had a major effect, because of the control that was exercised over the stages of joint manufacture. The variability sometimes observed is therefore difficult to explain. In the comparative studies reported here variability was minimised by using one mix of adhesive whenever possible, and control joints made with the standard etch pre-treatment were often included to give a comparative base. The differences and trends reported are therefore believed to be real.

3.2 Joint strengths obtained with simplified surface treatment and with contaminated surfaces

For rapid repairs in the field one cannot use the usual etch or anodic pre-treatments, and also it is likely that the surfaces will be contaminated with a variety of fluids and pastes used on aircraft. At best the surface pre-treatment will comprise solvent swabbing, abrasion and re-swabbing: at worst the solvent cleaning might be omitted, with only a dry wipe to remove most of the dirt, followed by abrasion. The results reported here distinguish between the effects of oil itself, use of abrasion instead of etching, and the combined effect of contamination with abraded surfaces. Most reliance was placed on peel load, because this is more sensitive than lap shear strength to variations in adhesion.

3.2.1 Effect of oil on peel load with etched adherends:
To determine the effect of oil alone adherends were first given the normal etch, then immersed in an aero engine oil, drained, and finally wiped dry to leave a thin film of oil on the surface (tab. 7). Comparative values for etched, clean surfaces were those obtained in examining the effect of temperature.

With the clean, etched adherends failure was almost entirely cohesive. With the oiled adherends failure was almost entirely in adhesion for adhesives H, I and M, as would be expected from their low strengths, whereas adhesives C, N, O and P failed largely in cohesion. The essential feature of these results is the wide diversity in oil tolerance shown by apparently similar adhesives.

Table 7:

Effect of oil on peel load obtained with etched adherends

Adhesive	Peel load (N/mm)			
	Etched only		Etched, oiled, wiped	
	Load	Standard deviation	Load	Standard deviation
C	8.9	0.20	6.7	0.67
N	7.6	0.37	7.1	0.48
O	8.9	0.61	6.1	0.79
P	7.9	0.37	6.6	0.25
H	8.2	0.31	Nil	–
I	8.5	0.22·	Nil	–
M	9.0	0.47	1.9	0.20

Note: L73 aluminium alloy, 3 day cure

3.2.2 <u>Comparison of peel load for etched and clean abraded adherends</u>: The
four adhesives (C, N, O, P) that bonded well to oily etched surfaces were next
tested for their ability to bond to cleaned, abraded adherends. To simulate
repair of a contaminated aircraft structure the adherends were first etched
and then immersed in mineral hydraulic oil, or aviation kerosine, or the
engine oil used before. The adherends were wiped dry, swabbed with
1,1,1-trichloroethane, hand-abraded with 240-grit paper and finally swabbed
again. The single control specimens were simply etched.

The results (tab. 8) show that abrasion gave much weaker joints than
etching. As might be expected, there was little or no difference between
contaminants, with the possible exceptions of Adhesives C and N, because only
traces of the original contamination would have remained after abrasion and
cleaning. With etched adherends failure was almost entirely cohesive. With
abraded adherends Adhesives C, O and P gave entirely adhesion failure, whereas
Adhesive N gave largely cohesive failure. Again, marked differences between
apparently similar adhesives are evident with N and P suffering less from
the change to abraded adherends.

3.2.3 <u>Comparison of strengths for etched, clean abraded and dirty abraded
adherends</u>: The earlier trials led to the choice of Adhesives O1 and P2 as
most likely to suit the intended use, and they were therefore evaluated in a
more rigorous simulation of contamination combined with limited surface
treatment. (Adhesive O1 is a development of O, said to have a higher hot
strength, and P2 is a slower curing variant of P. Both were used as pre-
mixes, whereas O and P were used in their separate-coat form.)

Mixtures were made to simulate 4 classes of contaminant likely
to be found on aircraft:

(i)   <u>Oils</u> - a mixture of 5 common oils, including a silicone oil;

(ii)   <u>Greases</u> - graphite and molybdenum disulphide greases;

(iii) <u>Corrosion preventives</u> - 3 types, including waxy or resin-based
thickeners and chromate inhibitor;

(iv)   <u>De-icing fluids</u> - 2 common types, based on glycols and alcohols

These mixtures were used either individually, or blended into a
"grand contaminant mixture" (GCM), and applied to vapour degreased adherends
before pre-treatment. The 6 pre-treatments compared were:

1:    etch (control joints, no contaminant);

2:    contaminate with GCM, dry wipe, swab, abrade, swab;

3-6:   contaminate separately with the 4 classes, dry wipe, abrade, dry wipe (ie no solvent cleaning).

The second represented a "good" field repair technique, the last four a minimal treatment that would leave traces of contaminant on the surface. In treatments 3-6 the contaminants were assessed separately to determine if any were particularly bad for adhesion.   Abrasion was done with a flap wheel and the solvent was paint thinner, as described in Sect. 2.2.2.   Cure was limited to a realistic 3 hours.   Results for peel load are shown in tab. 9 and shear strength in tab.10.

Table 8:

Comparison of etched with clean abraded surfaces, effect on peel load

Surface treatment	Contaminant	Peel load (N/mm)	Adhesive			
			C	N	O	P
Etch only	none	mean	7.5	6.7	10.2	8.1
		Standard deviation	0.50	0.28	0.37	0.22
Etched, oiled, cleaned abraded.	hydraulic oil	mean	2.8	5.6	3.5	5.1
		Standard deviation	0.24	0.26	0.21	0.18
	kerosine	mean	2.4	5.2	3.6	5.3
		Standard deviation	0.21	0.22	0.25	0.26
	lubricating oil	mean	2.2	4.4	3.5	5.1
		Standard deviation	0.25	0.26	0.31	0.28

Note: L73 aluminium alloy adherends, 3 day cure

131

Table 9:

Effect of abraded and contaminated surfaces on peel load.

| Pre-treatment | Contaminant | Peel load (N/mm) | | | |
| | | Adhesive 01 | | Adhesive P2 | |
		Load	Standard deviation	Load	Standard deviation
Etch	none	5.2	0.30	9.5	0.40
Swab, abrade swab	GCM	5.7	0.48	2.6	0.12
Dry wipe, abrade, dry wipe	oils	6.1	0.41	2.6	0.13
	greases	7.4	0.35	2.4	0.10
	corrosion inhibitors	6.5	0.11	1.8	0.11
	de-icing fluids	6.0	0.53	2.0	0.10

Note: L73 adherends, ∿ 3h cure

Table 10:

Effect of abraded and contaminated surfaces on lap shear strength

| Pre-treatment | Contaminant | Lap shear strength (MPa) | | | |
| | | Adhesive 01 | | Adhesive P2 | |
		Strength	Standard deviation	Strength	Standard deviation
Etch	none	22.0	0.60	41.8	0.87
Swab, abrade swab	GCM	17.0	1.10	30.2	1.52
Dry wipe, abrade, dry wipe	oils	17.7	1.09	28.4	0.30
	greases	19.2	0.59	31.0	0.89
	corrosion inhibitors	16.2	0.28	24.7	0.40
	de-icing fluids	16.3	0.68	29.2	1.12

Note: L73 adherends, ∿3h cure

It should be noted that Adhesive 01 after 3h cure is relatively undercured compared to P2, as will become evident from results to be presented in section 3.5. This probably accounts in part for its high peel strength with abraded surfaces, and relatively low lap shear strength. Compared to its predecessor (0) cured for 3 days (tab. 8) it gave lower peel strength with etched adherends, but higher with abraded surfaces. Adhesive P2 gave a similar peel strength on etched surfaces as its faster curing variant P, but lower strength on abraded surfaces. The composition changes introduced in variants 01 and P2 may have affected also the adhesion properties to poorly prepared surfaces.

The striking feature of these results is that solvent cleaning had little beneficial effect on initial strength, compared to simply dry wiping and abrading. Adhesives 01 and P2 appear to displace and absorb traces of a wide range of organic contaminants, including silicone oil, surface active agents, thickeners, that would ordinarily result in very weak adhesion. Indeed, a modern toughened room-temperature cure epoxy adhesive gave negligible peel strength even with the swabbed and abraded surface. Again, there were substantial differences in the strengths given by etched and abraded surfaces (with the exception of peel strength for Adhesive 01), as was also found in the earlier comparison (tab. 8). With the abrasion surface treatments the failure mode was mainly cohesive for the lap joints and with the peel joints made with Adhesive 01, whereas peel joints with P2 failed entirely in adhesion. With etched adherends failures were entirely cohesive.

3.3 Durability of joints in hot wet conditions

Most makers and published data on durability is for strength retention or stress rupture life of lap joints[2,3,5,12-17]. It was therefore of interest to apply the rapid but rigorous wedge cleavage test (Section 2.4.3), particularly with adherends having abraded and contaminated surfaces. The initial evaluation however used etched adherends (tab. 11), in order to give good adhesion and thus to emphasise the characteristics of the adhesive rather than the interface.

Table 11:

Fracture energy determined by the wedge cleavage test before
and after exposure to 95% rh at 50°C for 45h

| Adhesive | Fracture energy (kJ/m^2) | | | |
| | Unexposed | | After exposure | |
	Energy	Standard deviation	Energy	Standard deviation
C Set 1	3.0	0.14	1.3	0.27
C Set 2	3.3	0.48	1.7	0.59
H Set 1	3.6	0.44	2.5	0.31
H Set 2	4.0	0.75	3.1	0.71
I	2.0	0.40	1.6	0.23
N	2.8	0.18	1.1	0.05
O	2.6	0.28	1.2	0.24
P	3.2	0.50	1.9	0.36

Note: L73 adherends, etch pre-treatments, 3 day cure

For adhesives C and H two sets of joints were made at different times.
In both cases the second set comprised 12 joints and had thicker average glue
lines, 0.23 mm for sets 2 compared to 0.06-0.07 mm for sets 1.   Before
exposure all adhesives showed entirely cohesive failure, except M which gave
50-100% cohesive failure.   After exposure Adhesives M and O showed
substantial amounts of adhesion failure, and P gave 50-100% adhesion failure:
the remaining adhesives gave 80-100% cohesion failure.

The main feature of these results is that the toughness, durability
and failure mode for these room-temperature curing acrylates compare well
with equivalent values for modern, hot-cured, toughened, structural epoxy
adhesives used in aerospace bonding.   The fall in fracture energy on exposure
gave crack growths in the range 1.7 mm to 10.3 mm.

The effect of surface treatment was then examined for Adhesives 01 and
P2, using the same treatments and conditions as for the results shown in
tabs. 9 and 10, and the results are shown in tabs. 12 and 13.

Table 12:

Fracture energy determined by the wedge cleavage test before
and after exposure to 95% rh at 50°C for 45h: effect of
surface treatment for Adhesive 01

| Pre-treatment | Contaminant | Fracture energy (kJ/m^2) | | | |
| | | Unexposed | | After exposure | |
		Energy	Standard deviation	Energy	Standard deviation
Etch	none	2.7	0.11	1.6	0.10
Swab, abrade, swab	GCM	2.7	0.57	1.6	0.27
Dry wipe, abrade, dry wipe	oils	1.9	0.24	1.2	0.19
	greases	2.4	0.53	1.5	0.25
	corrosion inhibitors	2.3	0.36	1.5	0.37
	de-icing fluids	2.3	0.47	1.4	0.26

Note: L73 adherends, ∿3h cure

For Adhesive 01 the fracture energies are broadly similar to the
values given by its forerunner Adhesive 0 (tab. 11), and it is noteworthy that
contaminated surfaces gave values almost as high as etched or clean abraded
surfaces.   This is consistent with the high peel strengths shown by Adhesive 01
on contaminated surfaces (tab. 9), and as noted before this may be due in
part to a degree of undercure.   However, after the 2-day exposure at 50°C
the cure would have been largely complete with consequent reduction in
toughness of the adhesive, yet even after this exposure the fracture energies
still compared favourably with hot cure epoxy adhesives.   Also, during
exposure this adhesive gave substantial proportions of cohesive failure,
again indicating exceptional adhesion to poorly prepared surfaces.

Adhesive P2 behaved differently.   Fracture energies obtained with
etched adherends were broadly similar to those of the faster curing variant P
(tab. 11).   But with abraded or contaminated surfaces the fracture energies
were much lower, particularly after exposure, and traces of oil, greases and
corrosion inhibitors seemed particularly bad for durability.   This is

consistent with the differences in peel strength noted in tab. 9.   In contrast
to the initial strengths there is some benefit to be gained by solvent swabbing.
For this adhesive all crack growth during exposure was by adhesion failure,
regardless of surface treatment (initial cracks were cohesive).

Table 13:

Fracture energy determined by the wedge cleavage test before and
after exposure to 95% rh at 50°C for 45h: effect of surface
treatment for Adhesive P2

Pre-treatment	Contaminant	Fracture energy (kJ/m^2)			
		Unexposed		After exposure	
		Energy	Standard deviation	Energy	Standard deviation
Etch	none	3.3	0.11	1.7	0.09
Swab, abrade, swab	GCM	1.5	0.26	0.6	0.08
Dry wipe, abrade, dry wipe	oils	2.1	0.08	0.3	0.16
	greases	1.9	0.09	0.1	0.06
	corrosion inhibitors	1.2	0.03	0.2	0.17
	de-icing fluids	1.5	0.11	0.5	0.02

Note: L73 adherends, ∿3h cure

3.4 Effect of temperature and post-cure on joint strengths

Published and makers' data indicate considerable differences between
acrylate adhesives in their strength retention at high and low
temperatures[13-15,17] with some examples having good strengths at 100°C or
higher.   For the intended application usable strengths were required at
temperatures in the range -55°C and +120°C, so measurements were made at a
limited number of temperatures over that range.   The effect of a post-cure
of 1 hour at 80°C was also examined by determining the strength at 80°C only.
The results are shown in tab. 14 for lap shear strength and tab. 15 for peel
strength.   For each entry the 2 values are respectively strength and standard
deviation.

Table 14:

Effect of test temperature and post-cure on lap shear strength

Temperature (°C) Adhesive	Lap shear strength (MPa)					
	-55	RT	+80	+80 after post-cure	+100	+120
C (1)	19.1 3.63	28.8 2.08	12.9 0.91	15.3 1.03	- -	- -
H (1)	26.5 2.56	40.4 2.06	22.8 2.15	23.4 1.96	- -	- -
I (1)	28.4 2.03	37.2 1.71	24.8 2.44	23.7 1.97	- -	- -
N (1)	10.3 2.06	19.0 1.89	6.3 0.83	10.1 0.44	- -	- -
O (1)	18.5 4.48	26.6 2.05	9.2 0.27	10.8 0.53	- -	- -
P (1)	18.4 2.84	32.1 2.81	16.7 1.54	17.0 1.77	- -	- -
O1 (2)	19.9 0.74	17.0 1.09	- -	- -	- -	3.4 0.44
P2 (2)	17.0 1.45	30.2 1.52	10.8 0.30	- -	6.9 0.34	3.6 0.41

Note: L73 adherends; (1) etch pretreatment, 3 day cure; (2) swab, abrade, swab pretreatment, ∿3h cure

The lap shear strengths show considerable variation between adhesives, with H and I outstanding over the temperature range -55°C to +80°C.   Also, these 2 adhesives gave cohesive failure over the whole temperature range, whereas all the other adhesives gave entirely adhesion failure at -55°C and mainly cohesive failure at room temperature and above.   Only adhesives C and N show a substantial effect of post-curing, and the lack of effect with the others was perhaps because most were held at the 80°C test temperature for 15 minutes before testing, in accordance with specification requirements.   They thus received a substantial post-cure in any case, and to minimise this effect in later work Adhesives O1 and P2 were tested as soon as they reached test temperature.   This may account for P2 being substantially weaker at 80°C than P, whereas they are similar at room temperature and -55°C.   Note also

that for O1 and P2 the pretreatment and cure time differed.   The results for
Adhesive P2 show that strength fell sharply over the range 80-120°C, which
might be expected because the cured adhesives consist largely of polymethyl-
methacrylate with a Tg of 105°C.   Heat distortion temperature of an early
acrylate was found to be 102-104°C after post-cure at 80°C, and 96°C after
room temperature cure[18,19].   For comparison, makers' data for a range of room-
temperature curing epoxy adhesives show that most have strengths in the range
7-12 MPa at 80°C, and it was found that after room-temperature cure the heat
distortion temperatures were typically in the range 27-52°C[19].   Thus acrylate
adhesives cured at room temperature do show a potential for better performance
at moderately elevated temperature compared to room-temperature cured epoxy
adhesives.

Table 15:

Effect of test temperature and post-cure on peel load

Temperature (°C) Adhesive	Peel load (N/mm)			
	-55	RT	+80	+80 after post-cure
C	Nil —	8.9 0.20	4.9 0.27	5.2 0.52
H	1.4 0.18	8.2 0.31	9.2 0.27	9.1 0.31
I	3.3 0.16	8.5 0.22	11.9 0.44	11.7 0.28
N	Nil —	7.6 0.37	3.7 0.43	6.7 0.37
O	Nil —	8.9 0.61	5.6 0.32	5.6 0.48
P	Nil —	7.9 0.37	8.6 0.34	8.7 0.54

Note: L73 adherends, etch pretreatment, 3 day cure

The main feature of the peel load results is the low or zero values
at -55°C, presumably because the toughening rubbers used in these adhesives
had Tg values close to or above that temperature.   The results for Adhesives
H and I are consistent with the lap shear data in showing better strength
retention at the temperature extremes, and again only these 2 adhesives

gave cohesive failure at -55°C. Only Adhesive N shows any effect of post-cure, consistent with its behaviour in lap joints.

3.5 Effect of alloy composition on cure rate and strength

Transition metal ions affect the cure rate of acrylate adhesives in ways that are not well understood: in some cases cure is accelerated, in others retarded (by zinc for example[1]). Adherend surface composition may therefore modify cure, and work elsewhere showed wide variations between alloys in the cure rate and ultimate strength attained[13,14]. In the present work comparison was made between four common aerospace alloys of varying composition listed in tab. 16, which shows the nominal contents of alloying elements either in the alloy itself, or where appropriate in the surface cladding layer; the yield points are also listed because lap shear strength increases with increase in yield point[20], whereas peel load decreases[21].

Table 16:

Nominal content of alloying elements in contact with the adhesive

Alloy	Alloying element contant (%)*	Yield point (MPa)
BS L157 (bare)	Cu 4.4, Mn 0.8, Si 0.8 Mg 0.5	390
BS L165 (clad)	None	355
ASTM 7075-T6 bare	Cu 1.6, Mg 2.5, Zn 5.6, Cr 0.3	470
ASTM 7075-T6 clad	Zn 0.8	435

*Note: from published specifications, not actual surface analyses

To give a good simulation of real conditions, the degreased adherends were first covered with the "grand contaminant mixture" described in sect. 3.2.3 and then pretreated by the swab, abrade, swab sequence used before. Strengths were measured after cure times of about 2, 4 and 6 hours and 9-12 days at a temperature of 20 ± 2°C for Adhesives O1 and P2. Results are shown in figs. 3a and 3b for lap shear strength and figs. 4a and 4b for peel load. For clarity only those alloys showing major differences are included; scatter bars show standard deviation.

139

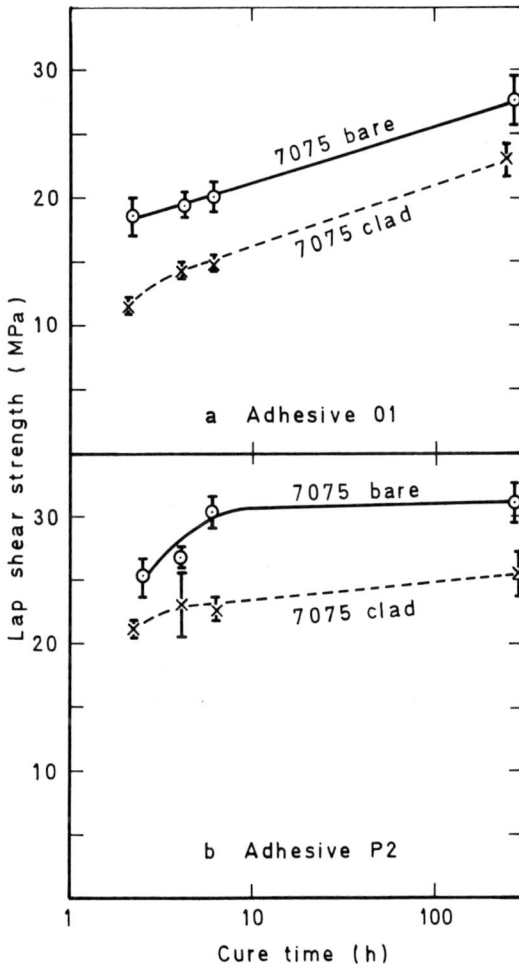

Fig. 3a & b Effect of cure time and alloy
on lap shear strength

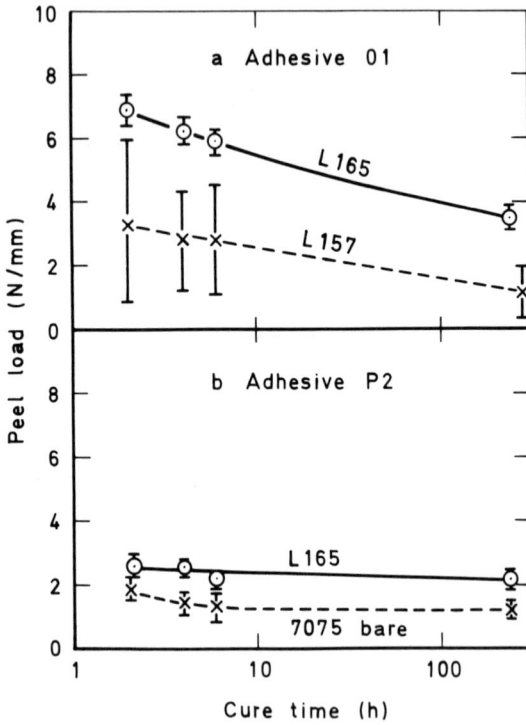

Fig. 4a & b Effect of cure time and alloy on peel strength

Comparing first the two adhesives, it is evident that although 01 hardens in less than 2h the cure then continues slowly for at least 10 days, giving higher lap shear strength and lower peel load; in contrast, cure of Adhesive P2 practically stops after 4-6h at 20°C.

Comparing the effects of alloy type, firstly for Adhesive 01 the largest difference in lap shear strength was between bare and clad 7075 (fig. 3a), with alloy L165 fairly similar to 7075 clad. Alloy L157 initially gave joint strengths as high as 7075 bare, but after 10 days cure the strengths were similar to those given by 7075 clad and L165. Alloy L165 gave the highest peel loads (fig. 4a), with 7075 bare slightly lower; and 7075 clad gave similar values after 2h cure but a faster decline in peel load with increased cure time. However, the major difference was found with alloy L157, which gave not only low peel loads but also very high variability. (A repeat set of joints gave somewhat higher loads but again high variability.) This scatter was both between specimens and along the length of specimens: in the worst cases peel load changed by up to 4.9 N/mm over a peel distance of 30 mm, and the difference between maximum and minimum points in a single trace ranged up to 6.4 N/mm. The strength variations were correlated with changes in mode of failure: higher load regions showed more cohesive failure with a thicker layer of adhesive remaining on the peel strip, whereas lower load areas gave more adhesion failure and a thinner residual layer. These contrasts of failure mode were similar to those occurring with increase in cure time for the other 3 alloys. They gave uniform modes of failure for any one set and cure time, but as the cure time increased the proportion of cohesive failure decreased from 100% to around 70% after 10 days cure, and the residual adhesive layer also became thinner. This was presumably an effect of decreasing toughness as cure progressed, rather than any change in adhesion, and the same explanation is believed to apply to the variability observed with alloy L157: ie the weaker regions were caused by areas of more fully cured, more brittle adhesive. Neither the lap shear joints with adhesive 01, nor either form of joint with Adhesive P2, gave high variability with alloy L157.

With Adhesive P2 the largest difference in lap shear strengths were again shown by the clad and bare versions of 7075 alloy (fig. 3b), with the other alloys giving values close to those of 7075 bare. Highest peel loads were given by alloy L165 and lowest by 7075 bare.

It should be noted that the strength differences are not simply explained by differences in yield points between alloys; (note that L165 is the clad

form of L157). Rather, it is thought that differences in surface composition
are the main cause. However, it is impossible to propose any detailed
explanation in the absence of data for actual composition at the surfaces, and
while lacking a good understanding of the effect of transition metal ions on the
adhesive cure chemistry. The extreme variability of peel strength for
Adhesive 01 with alloy L157 may have arisen from variations in surface
composition, caused by variations in depth of abrasion through a surface layer
that differed in composition from the bulk due to segregation of alloying
elements. The tendency for 7075 clad alloy to give lower strengths may be
attributed to the zinc in the cladding layer, but in contrast 7075 bare alloy
gave higher strengths despite a much higher zinc content. Again, surface
segregation may have given a composition differing considerably from the bulk,
and perhaps the other transition metals in 7075 bare outweigh the effect of
zinc. (Grit-blasted 7075 bare was found to have much less zinc at the surface
than implied by the nominal composition[22].)

It is concluded that alloy surface composition can affect adhesive cure
rate and joint strength. The surface composition in turn will depend on
bulk composition, metallurgical state and type of surface treatment. In
particular, abrasive surface treatments could give variable surface composition,
unless well-controlled.

4 CONCLUSIONS

Some potential causes of low and variable strengths with acrylate
adhesives are well known, such as too long an open time, too thick a glue line
or movement during cure. In the present work the effect of open time was found
to be not as serious as expected, particularly with some materials. Oxygen
inhibition and variations in atmospheric humidity did not cause variability.
Peel strength was very low with one adhesive if catalyst was applied to the
thin peel member, and variations in alloy surface composition gave variable
strength with one combination of alloy and adhesive. Apart from these
effects the observed variability was not thought in general to be a serious bar
to successful use of these materials.

Several adhesives gave reasonably strong and durable bonds to
contaminated surfaces prepared only by solvent cleaning and abrasion, although
this capability varied widely between materials. Even omission of solvent
cleaning did not greatly reduce strength in two cases. Some adhesives gave
high lap shear strengths at both -55°C and +80°C after only room-temperature

cure, but no adhesive was found that combined oil tolerance and high strength over a wide temperature range.   Peel strengths were low at -55°C, indicating brittleness at this temperature.   Joint strengths and cure rates depended to some degree on alloy surface composition.

The acrylate adhesives are thus a useful complement to the room temperature cure epoxies, but differ widely amongst themselves and have features quite distinct from the epoxies.   These characteristics must be appreciated if they are to be used effectively, and in particular it is important to test with the adherend material and surface condition that is of interest.

ACKNOWLEDGEMENTS

The following companies helped considerably by supplying much useful background information on their products, and in discussion: Agomet Klebstoffe, Bostik, Ciba-Geigy, Permabond Adhesives.

REFERENCES

1.   Harrison, K.W.   'Rapid acrylic adhesives', Symposium, 'Fastening with adhesives', Nat. Exhib. Centre, Birmingham, September 1978.

2.   Harrison, K.W.   'Acrylic engineering adhesives', Paint Manuf. and Resin News, vol. 51, no. 2, 23,25, 1981.

3.   Hauser, M. and Loft, J.T.   'Anaerobic and modified acrylic adhesives', Adhes. Age, vol. 23, no. 12, 21-24, 1980.

4.   Krebs, P.   'Basics and applications of methacrylate adhesives', Bond voor Materialenkennis, Kring Hechting, Int. Agrarisch Centrum, Wageningen, Holland, Nov. 1979.

5.   Lees, W.A.   'The science of acrylic adhesives', Brit. Polym. J., vol. 11, no. 2, 64-71, 1979.

6.   Prane, J.W.   'Reactive adhesives', Adhes. Age, vol. 23, no. 8, 35-37, 1980.

7.   Charnock, R.S.   'Heat resistant toughened adhesive composition', Europ. Pat. Appln., no. 0 034 046, Aug. 1981.

8.   Charnock, R.S.   'Polyisoprene toughened adhesive composition', Europ. Pat. Appln., no. 0 040 079, Nov. 1981.

9.   Charnock, R.S.   'Butadiene toughened adhesive composition', Europ. Pat. Appln., no. 0 044 166, Jan. 1982.

10.   Zalucha, D.J., Sexsmith, F.H., Hornaman, E.C. and Harris, T. 'Structural adhesive formulations', U.K. Pat., GB 2019415B, Oct. 1982.

11. Bowditch, M.R. and Stannard, K.J. 'Adhesive bonding of GRP', Composites, vol. 13, no. 3, 298-304, 1982.

12. Zalucha, D.J. 'New acrylics structurally bond unprepared metals', Adhes. Age, vol. 22, no. 2, 21-23, 1979.

13. Wilkinson, T.L. 'Acrylic adhesives: new way to bond aluminium auto-parts', Adhes. Age, vol. 21, no. 6, 20-23, 1978.

14. Wilkinson, T.L. and Tyler, D.P. 'Acrylics improve for bonding automotive aluminium alloys', Adhes. Age, vol. 24, no. 12, 34-38, 1981.

15. Coleman, J.F. 'Modified acrylic adhesives afford durable bonding', Adhes. Age, vol. 17, no. 2, 25-28, 1974.

16. Jacobi, C.H. 'Application of acrylic based room temperature curing adhesives to structural helicopter bondments', 19th Nat. Symp., Soc. Adv. Mat. and Process Engng., Apr. 1974.

17. Coleman, J.F. 'Durability studies on toughened acrylic resin adhesives', 5th Nat. Tech. Conf., Soc. Adv. Mat. and Process Engng., Oct. 1973.

18. Stone, M.H. 'The effect of adhesive properties on the low temperature strength of joints between metals and carbon fibre reinforced plastics', Int. J. Adhesion and Adhes., vol. 1, no. 4, 203-207, 1981.

19. Stone, M.H. Unpublished report, Royal Aircraft Establishment, 1979.

20. Adams, R.D. and Harris, J.A. 'Energy absorption and strength of adhesive lap joints under impact loading', 21st Ann. Conf. on Adhesion and Adhesives, City University, London, March 1983, Paper 1.

21. Crocombe, A.D. and Adams, R.D. 'An elasto-plastic investigation of the peel test', J. Adhesion, vol. 13, no. 3-4, 241-267, 1982.

22. Poole, P. Unpublished work, Royal Aircraft Establishment, 1983.

NOTATION

G = fracture energy

E = Young's modulus of adherends

d = displacement caused by the wedge

h = adherend thickness

a = crack length

Tg = glass transition temperature

Chapter 8

ADHESIVE BONDING IN THE AUTOMOTIVE INDUSTRY

E. D. LAWLEY

B.L. Technology Ltd, Cowley, Oxford, OX4 5NL

SYNOPSIS

The use of adhesives for mass production automotive applications has steadily increased over the last decade. Initially the adhesives were an extension of the low strength gap filling interweld sealers and contributed little if anything to the overall strength of the assembly. The adhesives now used are required, in many instances, to accommodate the accumulative errors in joint fits and must have a gap filling property. The adhesive must be designed to suit the production condition, which can lead to a variety of products being used on an assembly.

Adhesive bonding is now being designed into the vehicle from the concept stage and the requirements such as flange width, designed gaps, etc. are included. It has several advantages, the ability to join different materials, seals the joints and no heat distortion as experienced by welding methods.

In the U.K. automotive industry all adhesives were applied manually, which leads to significant quality variations due to operator performance. To overcome this most large companies are introducing automatic methods of application, both dedicated and robotic programmable systems. The European automotive industry has been using automatic application systems for several years for specific simple shapes but the more complex shapes are generally still manual.

1 MATERIALS AND DESIGN

During the last 10 to 15 years the use of non-metallic sealers and adhesives in the automotive industry has gradually increased. Developments have occurred in both the materials used and the methods of application.

The early materials were designed to fill gaps in assemblies to stop water, fume and dust leaks into the passenger compartments, and were used to replace rubber gaskets, brazing and soldering. At the time the demands on the material were minimal, the material had to be capable of filling the gaps and to remain in position during the service life of the vehicle.

Sealers were used during the initial body build between mating surfaces which were subsequently bolted or welded together.

The materials must :-
. Allow spot welding
. Not produce corrosive by-products
. Be compatible with the subsequent processes

If they inhibit spot welding or significantly reduce the weld strength they may only be used in areas where other jointing methods such as bolting/ riveting/stapling are specified.

Materials of this general type are still very much in evidence in the automotive body shell and the same constraints apply equally today as they did in the early days. The type of material selected is dependent upon the joint, its position and the subsequent process parameters.

Since these early beginnings the materials have been developed to give better/longer term performance. The concept of using non-metallics at the body build stage has been generally accepted by both production workers and the designer engineers.

From a design aspect it was soon realised that as well as sealing joints, materials of this type could be used to increase the joint strengths by replacing the sealers with adhesives and including them in the joint design. The adhesive is used to either supplement or replace the traditional jointing methods and to distribute the loads within the joint.

Joint designs have progressively changed to allow adhesive bonding to be used. Unless this is considered at the design/concept stage, it may be too late by the time it is in production to have flanges modified to suit a bonding condition.

It is important that in any design team the process engineering aspects are fully assessed and taken into account so that the process may be planned into the production facility with the knowledge that it can be achieved.

The materials used within the automotive industry, in the body shell, can be divided as follows:-

. Sealers   - to fill gaps
. Semi-structural adhesives - to work in conjunction with other fixing
                                 methods $(1N/mm^2)$
                              - to seal joints
                              - to bond inner and outer skins
                              - to bond low stressed assemblies
. Structural adhesives   - prime method of fixing
                              - high strengths $(15/25N/mm^2)$
                              - to give additional strengths where
                                 increasing the welds is not practical.

All these groups of materials need to be :

. Compatible with subsequent processes
. Non-toxic
. Non-corrosive
. Temperature stability
. Compatible with unclean substrates.

The formulation of the product is generally the result of close liaison with the supplier; the product may be based on a variety of systems such as PVC, epoxy, polyurethane, rubber, etc. or indeed a blend of certain base combinations.

As the demands on these materials increases with increasing design awareness further developments in the materials are inevitable but whatever the design demands, it is imperative that the correct shopfloor procedures are adhered to and, more important, the shopfloor is engineered to accommodate these methods.

2 CURING

In any mass production situation curing can be a problem. In order to allow the assembly sequence to take place the material must have a reasonably long time prior to any cure occurring to allow the additional operations to be completed without hindrance by the adhesive.

Curing is genrally achieved during the paint stoving cycle when temperatures up to $180/190^{o}C$ are achieved for periods up to 30 minutes. Using this technique the cure is achieved at no additional cost, but to do this the structure must be inherently stable in the pre-cure condition either through the design or by using a secondary fixing media such as welding, riveting, etc.

Alternatively, the curing can be carried out soon after assembly using either air circulation, radiant heaters or HF induction heating systems so that the assembly is stable and will not distort during the subsequent handling and processing.

The choice of curing systems will be dependent upon the production constraints, materials used, panel/assembly shapes. Within the auto industry a whole variety of curing systems are currently in use both in the UK and Europe. If a curing system is to be used on the sub-assemblies it is important that it is jigged so that thermal distortion does not occur, otherwise scrap assemblies will be produced.

3 THE FORM OF THE ADHESIVE

The bulk of the adhesives used are in the familiar extrudable form supplied in any type of container from the cardboard tube to the familiar 200 litre drums.

The materials can be single or multi-component products. In the case of the multi-component efficient mixing and purging systems must be used. Multi-component materials are rarely used in vehicle construction because the single component materials are more readily handled and are satisfactory for the application.

In some situations the extrudable form of adhesive is not acceptable and 'solid' adhesives are used either as a simple preformed strip or as a strip on a carrier of film or fabric. This form is less widely used and its use is restricted partly by the lack of suitable applicators and the limited range of materials available in a handleable form.

4 METHODS OF APPLICATION

The design engineers, usually in conjunction with the materials technologist, select the material to be used and based on the theoretical performance of the adhesive the strength of the structure is calculated.

The strengths achieved in production will only relate to the theoretical values if the correct quantity of adhesive is applied in the correct locations. It is these aspects which can give wide variations in final products.

Currently within the auto industry most sealants and adhesives are applied manually although some manufacturers are seriously considering automation for these tasks, to both relieve operator boredom, obtain a higher reproducible quality standard and improve the working environment.

5 MANUAL APPLICATION

As previously stated the bulk of the adhesives used are manually
applied and few developments have occurred in this area. The major developments
have been in the design of the extrusion nozzle and the guides fitted to the
nozzle to ensure 'accurate' location providing the nozzle is applied at the
correct attitude to the work-piece, leaning the nozzle one way or the other will
affect the positioning of the bead.

Any guidance system used must not hinder application, therefore, the
lengths of guides and their fixings can be important.

Another form of nozzle guide used by some companies is to form a
feature into the panel which will guide the nozzle and clearly identify those
areas to which the adhesive must be applied.

If two parallel seams close together are required then multi-nozzle guns
can be used to advantage.

One aspect often overlooked is the type of feed pipe to the extrusion
gun; until relatively recently swivel joints were not commonly used and it was
necessary to twist the pipe during application placing additional constraints
on the operator.

Since the introduction of the pipe line swivels the extrusion gun can be
twisted to any angle relative to the pipe without effort. This seemingly
obvious change enables a much better application to be achieved, but the
equipment must be maintained otherwise swivels will seize and the advantages
will be lost.

6 AUTOMATIC APPLICATION

There are two basic forms of automatic application :
- Dedicated systems designed and built to handle one component/
  assembly.
- Programmable systems which include the now familiar industrial
  robots.

Both systems are in use in Europe, Japan and the UK. The choice between
one system and the other is dependent upon predicted model life, complexity of
the component and frequency of engineering changes, cost effectiveness and
many other contributory factors.

The use of automated systems highlights a number of problems not present with manual operation in which the operator quickly adapts to a changed situation.

These may be summarised as :
. Accurate jigging of panels
. Consistent supply to application nozzle
. Automatic barrel change-over
. Consistent viscosity
. Freedom from air in the adhesive
. Controlled dispensing rate
. Regulated flow - no surging
. Clean material free from debris
. Temperature insensitivity

Changes in any of these properties will affect the quantity dispensed and variations will adversely affect the product and the process.

Assuming these features are controlled to within acceptable limits it is necessary to control the automated system. Assuming the dispensing flow rate is constant then in order to deposit a bead of constant dimensions the nozzle movement must be constant. With some systems this is difficult particularly when completing complex manoeuvres.

However, attention to the control systems, and in the case of some robots to the detail programming, can easily overcome this problem.

Within our own organisation the use of automated systems has highlighted the need for first class reliable backup equipment in order that the main system can operate consistently throughout the working day.

The automated application of adhesives is being closely evaluated by a number of auto manufacturers, if and when these are introduced then serious developments must be undertaken by the "glue handling companies" to develop sensing systems capable of sensing the low-flow rates involved to determine when blockages occur, so that through simple feedback loops the blockages can be automatically cleared without distruption to the production process.

7 TYPICAL APPLICATIONS

7.1 Adhesives

Door inner assembly to outer panel)
Bonnet inner to outer   )   non-stressed
Trunk lid inner to outer   )

Suspension points   )   in conjunction)
Main body frame joints)   with welding   )   stressed areas

7.2 <u>Sealers</u>

Bolt together joint faces
External body seams - interweld/inter flange
                         - external to the joint
Engine mating faces - liquid gasket

8   THE ADVANTAGES OF BONDING

The use of adhesives increases the joint strength, distributes the loads more evenly and in many instances this enables the alternative jointing methods to be reduced or eliminated.

Within the car industry the use of adhesives in the bolt on assemblies such as doors, bonnet, trunk lid has enabled the number of spot welds on the periphery to be reduced by more than 50% whilst retaining the original torsional strength of the assemblies.

With reduced welding there is less rectification required due to electrode damage, giving reduced manufacturing costs and improvements in visual quality.

The adhesive in the joints also imparts sealing, giving an additional benefit in terms of the reduction of water leaks, dust and fume leaks and corrosion.

Using bonding as a jointing media has enabled assemblies to be joined together even where it is impossible to gain access to either side of the joint, thereby increasing the scope of the design engineer.

Dissimilar materials e.g. steel, aluminium, plastics, glass, etc. may all be joined together by bonding. In the past it would have been necessary to use mechanical fixing methods which are visually unacceptable.

9   THE PROBLEMS ASSOCIATED WITH BONDING

The problems may be summarised as follows :

. Control of application
. Selection of the right material for the job
. Assembly fixture contamination
. Additional facilities if extra curing required
. Control of proportioning with multi-component materials

. Equipment maintenance

. Adequate process control

10 CONCLUSION

Adhesive bonding is a viable technique for joining components in a
mass production environment. It is necessary to select the right material for
the job and to control the process within the specified limits. Using
automatic application and control methods it is a very reliable technique.

Chapter 9

INTERFACIAL PROPERTIES OF IONOMER CEMENTS AND THEIR EFFECTS UPON MECHANICAL
PERFORMANCE.

K.A. HODD & M.J. READ*
Department of Non-Metallic Materials, Brunel University, Uxbridge

ABSTRACT

This paper compares the mechanical and hydrothermal ageing properties of ionomer
cements with that of conventional glass - thermosetting resin composites. It
discusses their performance in relation to the adhesion at the glass/polymer
interface and shows that the superior properties of the ionomer cements is due
to primary interfacial bond formation.

1. INTRODUCTION

Ionomer cements are particulate composite materials which have an ion-leachable
glass powder dispersed in a crosslinked polyacrylate matrix. These materials
display a number of unusual characterisitcs which are a consequence of a stable
interfacial bond between the glass powder and the polyacrylate matrix, in
particular ionomer cements retain their mechanical properties in environments
of high humidity. This paper compares the changes in mechanical properties
during accelerated hydrothermal ageing between two ionomer cements and two
conventional thermoset-particulate glass composites. It discusses the differences in relation to the glass/polymer interfacial bond. But, firstly an
introduction to the technology of ionomer cements is given.

Ionomer Cements

The antecedence of ionomer cement technology lies in dentistry with the
development of zinc polycarboxylate cements of Smith (1) and the ASPA
(AluminoSilicate PolyAcrylate) cements of Wilson and Kent (2). The ASPA
system has been extensively studied (3) and is one of the materials used in
this study; it is also commercially the most significant of the ionomer cements.
Ionomer cements are formed at ambient temperatures by reacting ion-leachable
inorganic powders with an aqueous polyacid; normally polyacrylic acid.
Although specially formulated multiphase glasses are used in dentistry,
ionomer cements can also be prepared with certain metal oxide and certain
minerals.

*Now at Raychem Corporate Laboratories, Swindon.

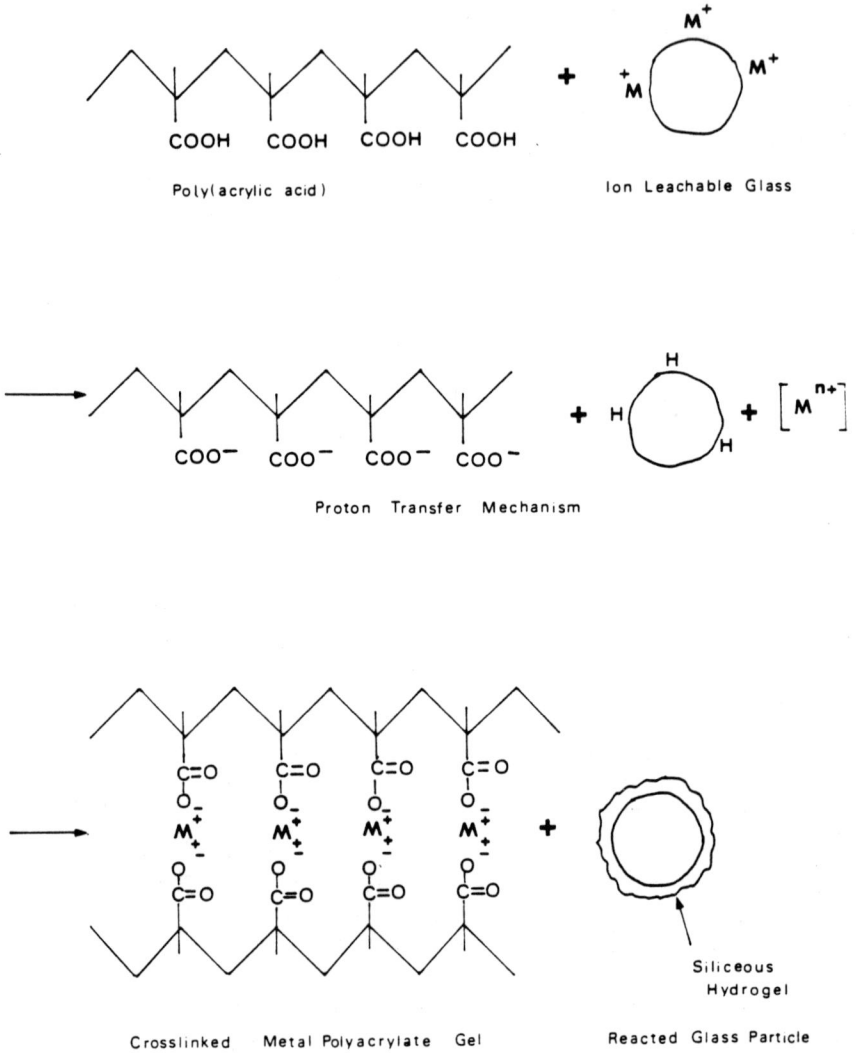

154

Figure 1. Schematic Representation of the Curing Mechanism of Ionomer Cements

Poly(acrylic acid)

Ion Leachable Glass

Proton Transfer Mechanism

Crosslinked Metal Polyacrylate Gel

Reacted Glass Particle

Siliceous
Hydrogel

The proposed setting mechanism of these materials, as shown in Figure 1, is an acid base reaction in which the cement sets by salt formation. Upon mixing the carboxylic acid groups (COOH) of the polyacid are ionised to carboxylate (COO⁻) groups and the freed hydrogen protons transfer with metallic ions from the glass surface. The liberated metallic cations react with the carboxylate groups to form various salts, including metallic ion crosslinks. It has been proposed that this setting mechanism leaves the glass surface depleated of functional cations and completely sheathed in a siliceous hydrogel (3,4). With the ASPA system an ion-leachable multiphase glass is employed which releases both Ca^{2+} and Al^{3+} ions. The Ca^{2+} ions are known to be readily leached from a dispersed phase of calcuim fluoride droplets on the glass powder surface and is responsible for the rapid setting characteristics required for dentistry. The aluminium polyacrylate is formed at a slower rate, as these ions are firmly embedded in the main phase of the glass' microstructure, and is considered responsible for the high mechanical strength and modulus of these materials which develops as a function of time as shown in Figure 2. The set materials are normally brittle in failure and display high strength and modulus when exposed to environments of high humidity, as shown in Figure 3. However, the ASPA materials are known to contain about 20% by weight of water which is mostly tightly held within its structure by co-ordination to metallic ions and to carboxylate groups of the polyacid. In dryer atmospheres some loosely bound water is lost, as the cement obtains a hydrometric equilibrium with its environment, and volumetric shrinkage occurs. With mild shrinkage (less than 2% at 60-80% RH) the mechanical properties improve due to a general tightening of the composites' structure. In more arid environments greater shrinkage occurs and the mechanical properties rapidly degrade due to interparticle contacts (4,5).

Consequently, ionomer cements have good ageing characteristics in humid environments. In contrast traditional composite materials, such as glass/thermosetting resin/coupling agent/systems, are known to be unstable in such environments as demonstrated by the typical accelerated ageing data of Scola (6) in Figure 4. This behaviour has generally been attributed to plasticisation of the matrix resin and, more importantly, to displacement of the resin from the interface by diffusion water (7-10).

156

Figure 2. The Ageing Properties of ASPA Cements (4,5)

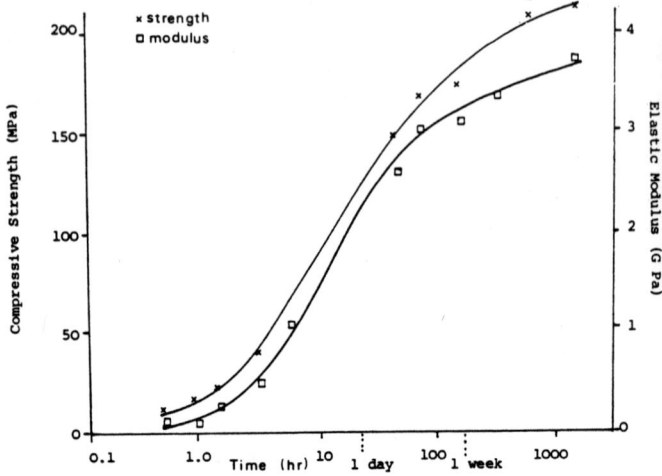

Figure 3. The Mechanical Properties of ASPA Cements (4,5)

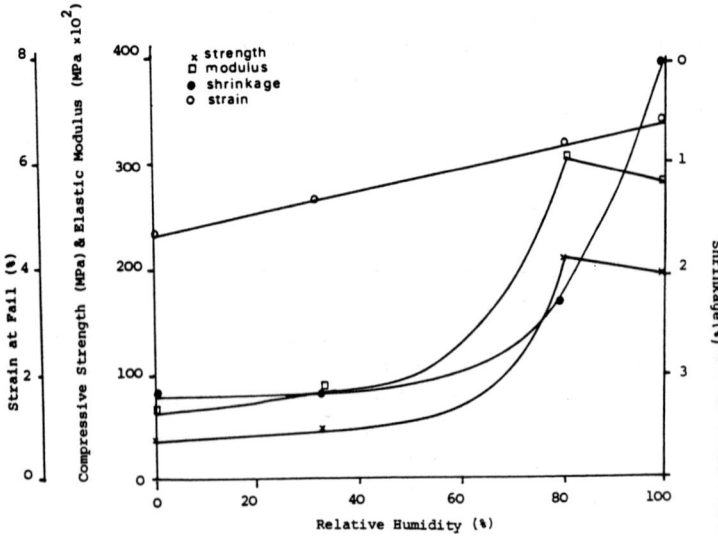

Figure 4. Loss of Mechanical Properties in Fibre Composites on Accelerated Ageing. After Scola (6)

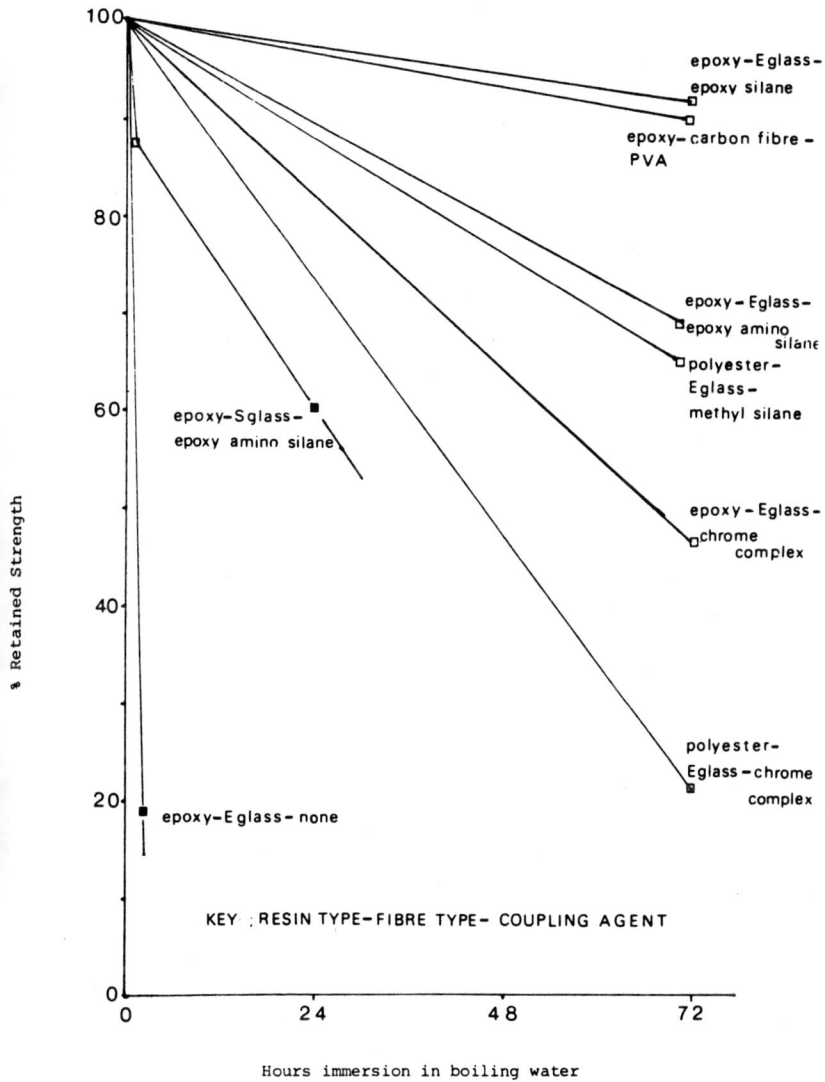

KEY : RESIN TYPE-FIBRE TYPE- COUPLING AGENT

Hours immersion in boiling water

2. EXPERIMENTAL

2.1 Ionomer Cements

Two ion-leachable glass powders were used in the preparation of the ionomer cements, these were:-

i) ASPA glass (G200), supplied by Amalgamated Dental International Ltd.

ii) BG glass, an experimental material developed at Brunel University.

Both had mean particle sizes of 10 microns with maximum size of 25 microns (Coulter Counter). The ionomer cements were prepared by reacting these glasses with a 50% aqueous polyacrylic acid solution, prepared from the anhydrous material (DP 1385) as supplied by Allied Colloids.

2.2 Thermosetting Resin/Glass Composites

The epoxy resin was a Ciba Geigy Araldite (liquid DGEBA) cured with a liquid polyamide. The polyester resin was Strand Glass Resin A in styrene cured with 1.0 ml of MEK peroxide per 100 g of resin. The BG glass powder was used with both these resin systems.

2.3 Composite Preparation and Testing

All the composites were mixed by hand spatulation at a volume fraction of 0.42 glass. The mixtures were cast into cylindrical moulds (60 mm diameter, 12 mm height) and cured at ambient temperatures in a desiccator at 66% RH for 7 days. The composites were hydrothermally aged by boiling in water for 24 and 48 hours. The mechanical properties were measured in compression on an Instron at a crosshead speed of 5.08 mm/min.

3. DISCUSSION OF RESULTS

3.1 Hydrothermal Ageing

The mechanical properties of the composites during accelerated ageing are shown in Figures 5, 6 and 7. The ionomer cements were mechanically superior to the conventional composites both prior to and after hydrothermal ageing. For example, after 48 hours ageing the conventional composites suffered a 50% modulus loss whilst the ionomer cements retained this property; in particular the BG cement performed well and was 300% stiffer and 50% stronger than the polyester composite after ageing. The differences in mechanical behaviour between these classes of composites were attributed to plasticisation of the matrix materials by water and to the nature of adhesion at the glass/polymer

159

Figure 5. Compressive Strengths of Composites with Hydrothermal Ageing

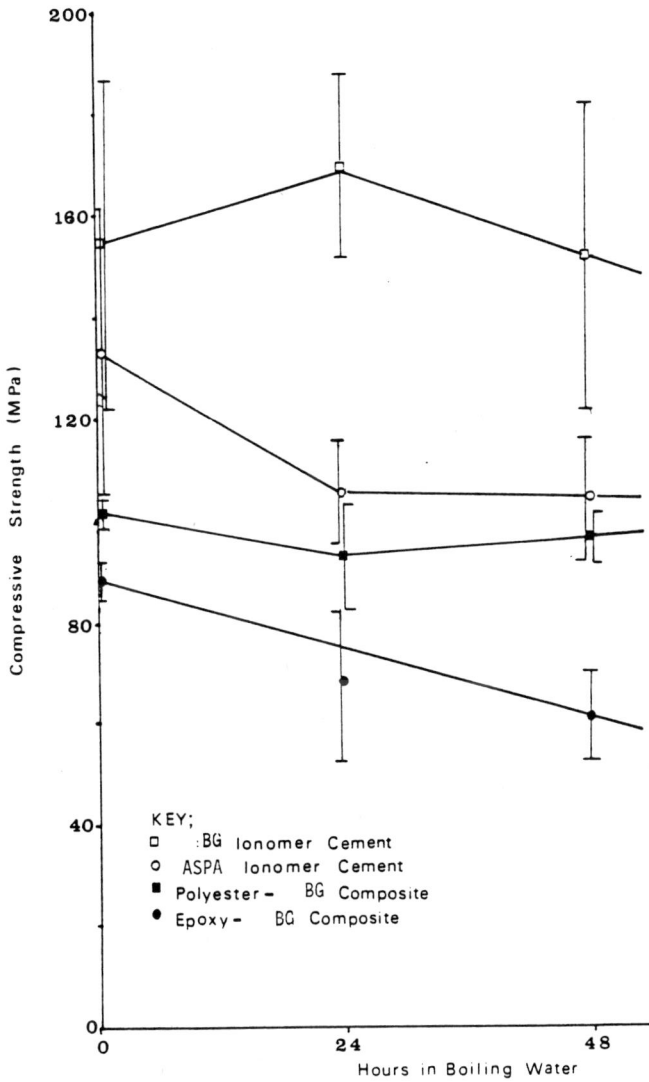

Figure 6.   Compressive Moduli of Composites with Hydrothermal Ageing

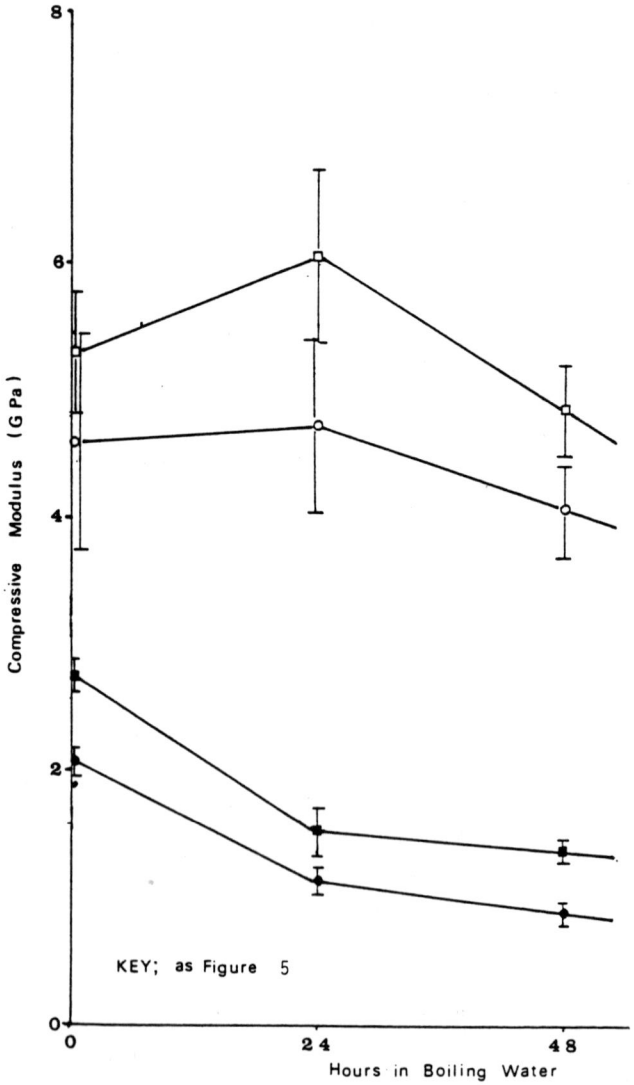

KEY; as Figure 5

Compressive Modulus (G Pa)

Hours in Boiling Water

interface. Water is known to diffuse into and plasticise both epoxy and
polyester resins (11), as is evident by the progressive increase in the strain
to fail with hydrothermal ageing (Figure 7). The ionomer cements were not
plasticised in this manner and the strain to fail remained stable during ageing.
With conventional composites the glass/resin interfacial bond is thermodynamic-
ally unstable and the resin can be displaced from the glass surface by diffusing
water (7 -11), which destroys the interfacial bond and reduces the mechanical
properties. However, this is a reversible process and properties are recovered
upon drying. Such behaviour is indicative of interfaces where secondary
bonding forces exist, such as Van Der Waals forces or Hydrogen bonding forces.
Hydrolytic stability is improved in systems where primary interfacial bonding,
such as covalent or ionic bonds, dominate. Consequently, the superior ageing
characteristics of the ionomer cements suggests primary interfacial bonding, ·
with the carboxylate groups from the polyacid forming stable metal-polyacrylate
bonds with ions firmly embedded on the glass particle surface as indicated in
Figure 8. Studies of the ASPA glass particle surface by ESCA have shown
that after acid attack metallic ions remain at the glass surface and are
capable of further reaction (12). This demonstrates that during the setting
of ionomer cements the glass is not depleted of functional cations nor completely sheathed in a siliceous hydrogel as had previously been believed (3). Further,
the hydrogel concept failed to explain the mechanical properties of these
materials as it would act as a weak boundary layer and would be unstable in
aqueous environment causing deterioration of the composites with hydrothermal
ageing. Thus, an alternative setting mechanism is proposed to explain the
properties of these ionomer cements and to correlate their properties with the
surface chemical composition of the ion-leachable glasses.

The surface chemical composition of the ion-leachable glasses were determined
by ESCA (Table 1). The BG glass had a higher aluminium content than the ASPA
glass which, in turn, was rich in calcium. The microstructure of such multiphase glasses shows that aluminium ions are embedded in the continuous phase
but calcium ions exist in a dispersed phase of fluorite (3). During setting
the dispersed phase is attacked to liberate Ca^{2+} ions which crosslink the
polyacid whilst the aluminium ions form reactive sites for primary interfacial
bonding, as depicted in the simplified model of Figure 9. The aluminium-
acrylate bond is also known to be hydrolytically more stable and mechanically

162

Figure 7. Strain at Fail Properties of Composites with Hydrothermal Ageing

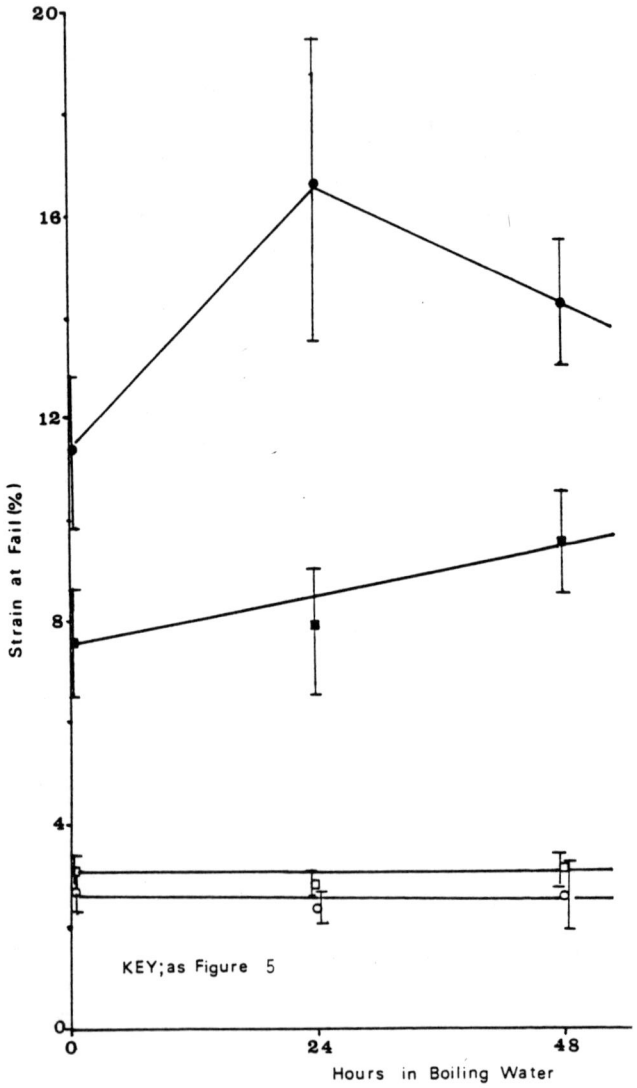

KEY;as Figure 5

163

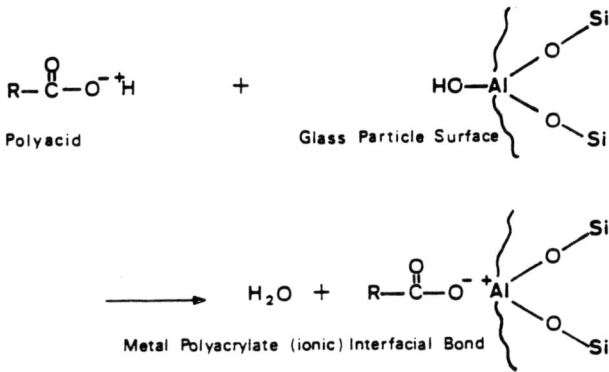

Figure 8. Interfacial Bonding Mechanisms in Ionomer Cements

superior to the calcuim-acrylate bond as it has a higher degree of covalency.
The calcium-acrylate bond is believed to be purely ionic (3). It is the
greater extent of aluminium polyacrylate/interfacial bond formation which
accounts for the superior mechancial and ageing properties of the BG cements.
The improved interfacial bonding of the BG cement was also found to produce
greater stability in dry environments, as shown in Figure 10.

Table 1.   ESCA Data of Ion Leachable Glass Surfaces

Kinetic Energy (e.v.)	Designation	Ratios of Peak Heights BG/ASPA Glasses
650	F Auger	0.53
795	F 1s	0.50
1045	Ca 2s	0.42
1134	Ca 2p 1/2	0.44
1138	Ca 2p 3/2	0.46
1333	Si 2s	2.11
1360	Al 2s	2.31
1375	Si 2p's	2.31
1405	Al 2p's	1.50

Figure 9. Proposed Mechanism for the Setting of Ionomer Cements

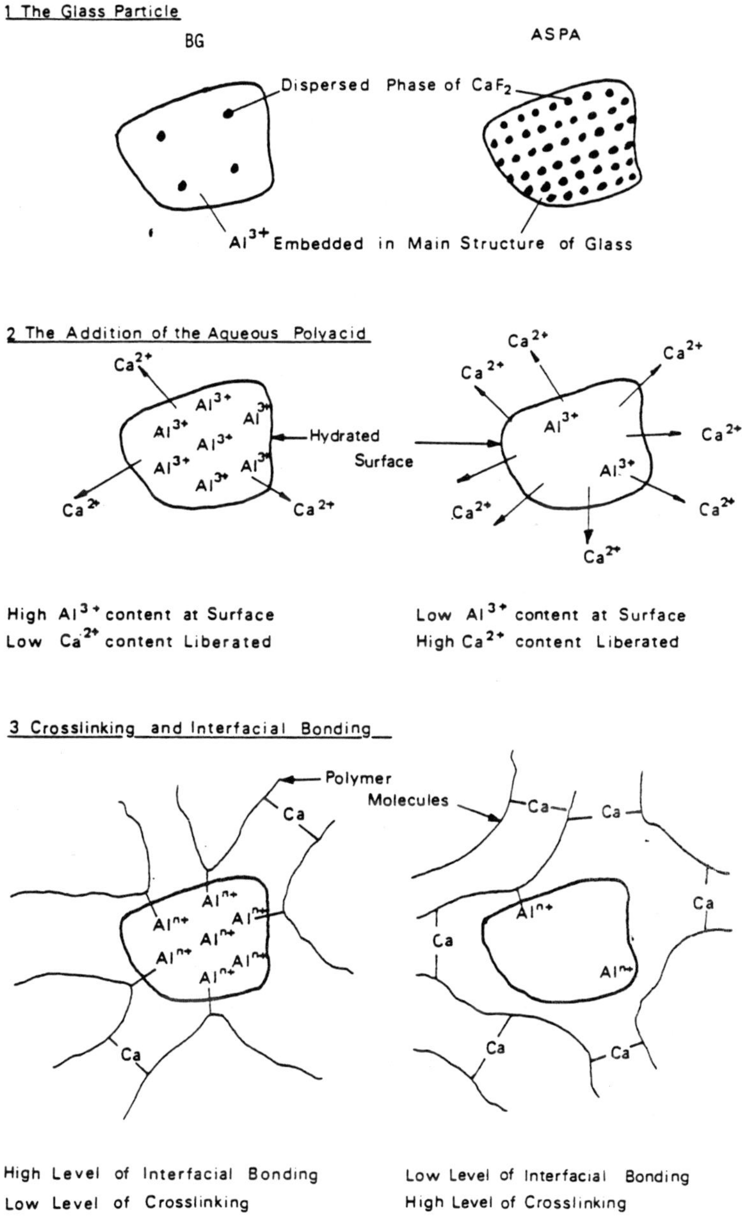
1 The Glass Particle
BG
ASPA
Dispersed Phase of CaF_2
Al^{3+} Embedded in Main Structure of Glass
2 The Addition of the Aqueous Polyacid
Ca^{2+}
Al^{3+}
Hydrated Surface
High Al^{3+} content at Surface
Low Ca^{2+} content Liberated
Low Al^{3+} content at Surface
High Ca^{2+} content Liberated
3 Crosslinking and Interfacial Bonding
Polymer Molecules
Ca
Al^{n+}
High Level of Interfacial Bonding
Low Level of Crosslinking
Low Level of Interfacial Bonding
High Level of Crosslinking

Figure 10. Mechanical Properties of ASPA and BG Cements Stored at Various Humidities

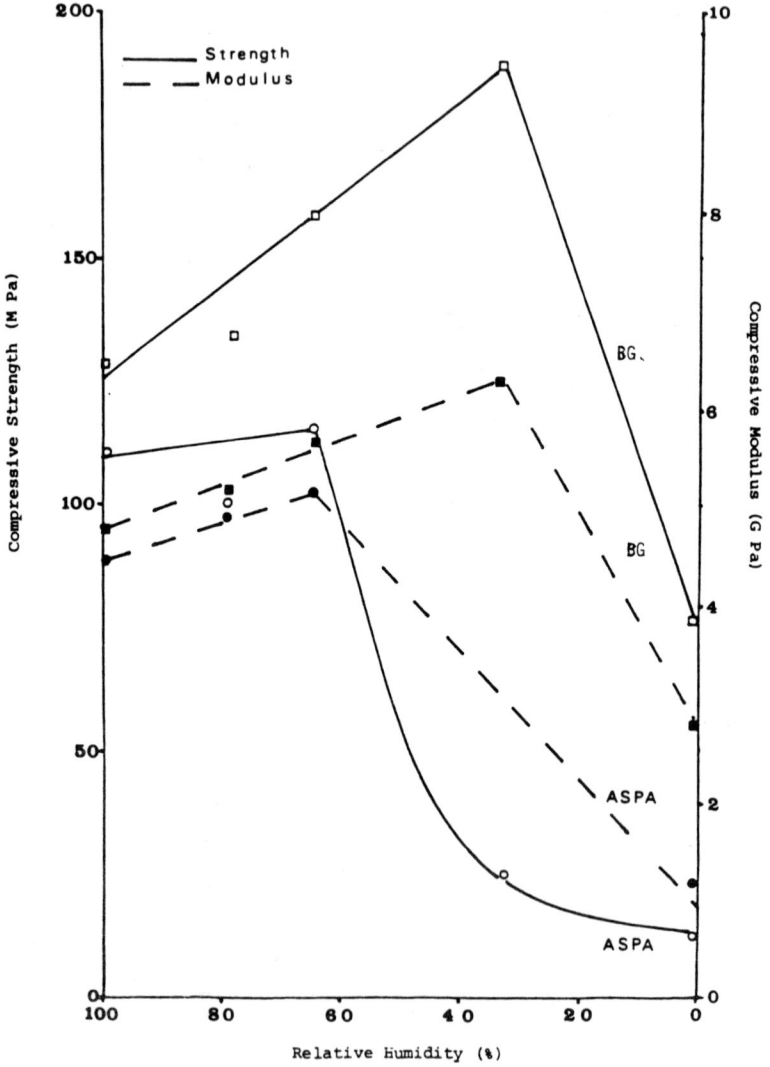

N.B. Cement Age 1 Month

4. CONCLUSIONS

Ionomer cements have superior hydrothermal ageing properties to conventional composite materials due to stable interfacial bond formation between the polyacid matrix and surface metallic ions at the glass powder surface. The BG cement was found to have a greater extent of aluminium polyacrylate formation than the ASPA cement which produced further improvements in mechanical performance and potential for use in dry environments.

5. ACKNOWLEDGEMENTS

The authors are grateful to Dr. W.G. Davies and to Dr. G. Bye of Blue Circle Technical Ltd. for their helpful discussions during the course of this study and to Blue Circle Technical for industrial sponsorship for MJR. The authors are indebted to Miss. P. Clothier for typing and preparation of this paper.

6. REFERENCES

1. D.C. Smith. Brit. Dental J. 125, 381, 1968.
2. A.D. Wilson, B.E. Kent. Brit. Pat., 1316129, 1969.
3. A.D. Wilson, S. Crisp; Chap. 4, Organolithic Macromolecular Materials, Applied Science 1977.
4. J.H. Elliot, L. Holliday, P.R. Hornsby; Br. Pol. J., 7, 297, 1975.
5. P.R. Hornsby; PhD Thesis, Brunel University, 1977.
6. D.A. Scola; Chap. 7, Composite Materials 6, Academic Press 1974.
7. W.C. Wake; Aspects of Adhesion, 7,232, Transcripta Books 1973.
8. R.H. Norman; Aspects of Adhesion, 8,267, Transcripta Books 1974.
9. N. Friend; p. 813; Mechanics of Composite Materials (Ed. Wendt). Pergaman Press 1970.
10. J. Conyn, D.M. Brewis, R.J.A. Salashi and J.L. Tegg, Adhesion 3, p. 13, Applied Science, 1978,
11. W.C. Wake; Adhesion and the Formulation of Adhesives, Applied Science, 1976.
12. K.A. Hodd, M.J. Read, Unpublished data, Brunel University, 1981.

Chapter 10

ADHESION TO GOLD

J. COGNARD and C. BOICHARD

Asulab S.A. - CH-2000 Neuchatel, Switzerland

1. INTRODUCTION

The object of this work of applied research is to compare the energy of adhesion to gold calculated from gold surface tension with the experimental strength of adhesive joints between two gold surfaces.

It seems widely admitted that the strength, W, of an adhesive joint is a function of the energy of adhesion, Wa, as expressed by the equation (1)

$$W = Wa + \phi \ Wa \qquad \qquad \dots\dots 1$$

where the coefficient ϕ is a characteristic of the adhesive depending only upon the adhesive properties and the conditions of stress (speed and temperature) (1).

Thus knowledge of Wa of given liquids on gold would allow comparison with results previously obtained with other substrates and known adhesives.

The energy of adhesion between two substrates is calculated from their surface tension. T. Smith (2) thought to close a long controversy as to whether gold was or was not a hydrophobic surface by experiments conducted on gold surfaces formed in ultra high vacuum. His conclusions, supported by more recent work (3) (4), is that "fresh" gold surfaces are hydrophilic and after some minutes the surface turns hydrophobic as previously observed by A. Zisman (5). In practical cases water does not wet the gold surface (6). We have thought that the bare solid surface escaped previous studies of wettability and that the use of recent theoretical works (7) (8) (9) could clear up the understanding of adhesion to gold.

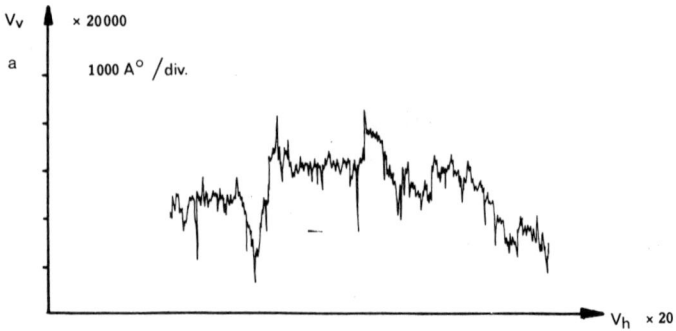

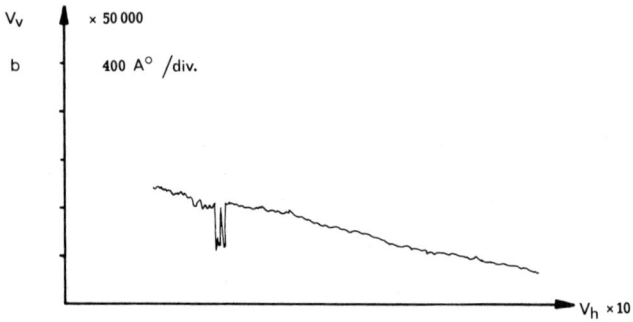

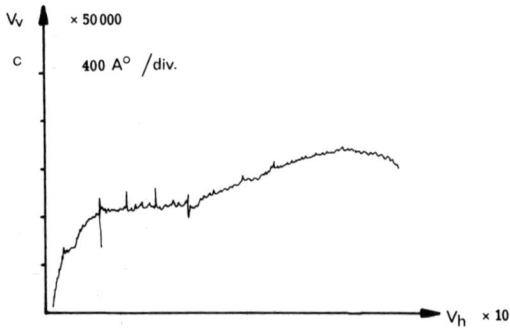

Fig. 1.   Talysurf plotting of the surface of one test specimen
    1a   Gold plated brass Ra = 0.17
    1b   Evaporated layer of gold on a flat glass, Ra = 0.013
    1c   Molded 24 ct. gold. Ra = 0.008
              The figures along the two axis give the magnification.

171

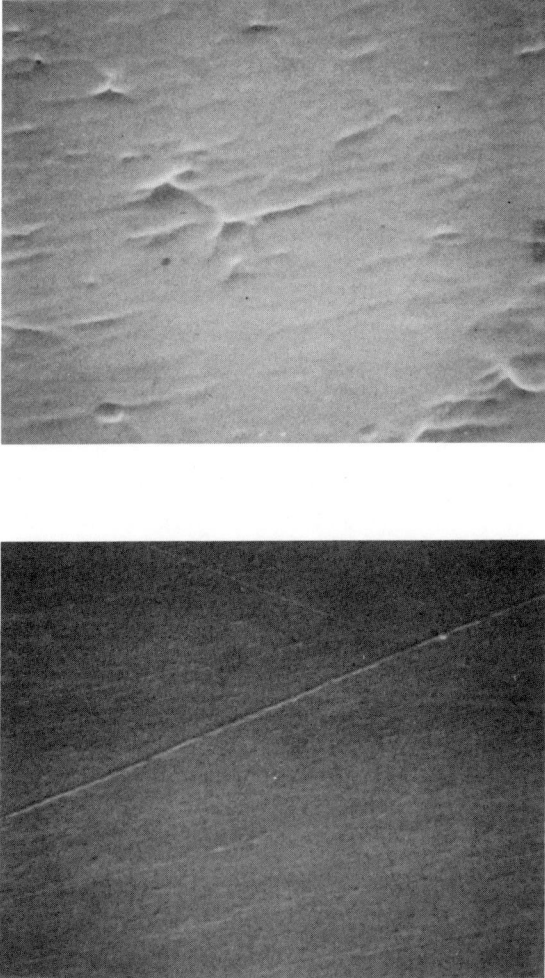

Fig. 2. Observation of the gold plated (above) and evaporated (below) surfaces by scanning election microscopy at 45°.

In a first part we will present measurements of contact angles and their interpretation, then in a second part will give results of the strength of adhesive joints between gold surfaces.

2.   EXPERIMENTAL

2.1.   Surface preparation

The different surfaces tested have been carefully considered as this is relevant to practical applications.

For the same reason our experiments are made at ambient conditions. Wettability studies are, except in paragraph 5, made on pure gold (99.999%) evaporated on a flat glass plate. The layer, 2000 Å thick to avoid any pin holes, is bound to the substrate through a first 500 Å chromium layer and a second 1000 Å thick layer of nickel.

For the purpose of comparison 18 carat gold-silver alloy polished with 1μ alumina powder, polished brass galvanically plated with 10μ gold (in fact 8μ Au-Co plus 2μgold cadmium (22ct)), and flat molded 24ct gold were also considered.

The state of the surface has been characterised by a Talysurf plot shown in fig. 1 and S.E.M. observations, some of them being shown in fig.2. Both show that the surface is very smooth. On evaporated layers the surface roughness average Ra is 0.003 (Ra is the arithmetical average of the departure of the profile above and below the reference line through the registered sample length). Gold plated samples are rougher but this does not show in the contact angles values.

2.2.   Contact Angles Measurements

Advancing contact angles were measured at ambient conditions on a magnifying projector. From the measurement of the drop diameter, l, and height, h, was obtained from the relation:

$$\text{tg}\,\frac{\theta}{2} = \frac{2h}{l}$$

The precision on θ is $\frac{\Delta\theta}{\theta} = \frac{\Delta h}{h} + \frac{\Delta l}{l} = 4\%$

Then from $\frac{\Delta\cos\theta}{\cos\theta} = \sin\theta\frac{\Delta\theta}{\theta}$ our precision is at least, 4% on the highest angles. This allows one to obtain the spreading coefficient $S = W_a - W_c$ with a precision better than 8%. For each contact angles measurement seven drops of 0.5μl were measured for the same solvent.

Liquids are of the purest grade but their surface tensions were period-
ically checked.

2.3. Surface Tension

The surface tension of the liquids were measured with a 9.5mm platinum
ring on a "Kruss"* tensiometer with automatic correctiion. The dispersive
components of the surface tensions were obtained from the contact angles on
polythene or paraffin cast over a glass plate. All values are given in mJM^{-2}.

2.4. Cleaning Procedures

The evaporated specimen were handled with the techniques used in the
electronic industry to prevent surface contamination.

Measurements were made under cover, the specimen stored in Teflon baskets
and transported in "wafer carrier storage boxes". Text-books recommend a
solvent degreasing.

Three cleaning procedures were compared: a tri-chlorethylene rinse
followed by an isopropanol rinse and hot air drying; or boiling sulfuric acid,
D.I. water rinse and spin drying; and detergent cleaning in an ultrasonic bath
followed by D.I. water rinse and hot air drying. Wettability studies indicate
that all surfaces have the same energy but joint strength is higher on detergent
cleaned surfaces than on those that have been cleaned with a solvent. Then we
retained the following procedure**: dip in a hot ($70^{o}C$) solution of a detergent
known to give film free surfaces under ultra-sonic agitation for ten minutes
and either D.I. water rinse and spin drying (evaporated plates) or three isopropanol rinses followed by hot air drying (galvanically plated samples); both
processes giving the same results.

2.5. Silanation

Clean, dry surfaces are silylated by dipping in a 5% solution of silane in
methanol which is aged for 24 hours. The test specimen are then rinsed with isopropanol and dried in a hot-air gun. The silanes are obtained from Dynamit-Nobel
AG

* A. Krüss D2000 Hamburg 60 Postfach 605265

** After presentation of this work some doubts were expressed concerning the use
   of detergent in wettability studies. We would like to insist on the point that
only the joint strength and not the critical tension value is changed when using
a non-filming detergent.

SOLVENTS	γ_L	γ_L^d	$\theta°$	Cos θ	γ_S	W_a	W_a^d	W_a^p
N. Hexane	18.4	18.4	0	1	> γ_L	-	-	-
P. Methylsiloxanne(1)								
47V3	19.2(1)	17.2(2)	0	1	"	-	-	-
47V20	20.6(1)	18.6(2)	0	1	"	-	-	-
47V100	20.9(1)	18.9(2)	0	1	"	-	-	-
P. Methylphenylsiloxanne	21(1)	19(2)	0	1	"	-	-	-
P. Methylsiloxanne 47V500	21.1(1)	19.1(2)	0	1	"	-	-	-
I. Butanol	23.7		0	1	"	-	-	-
N. Octane	26.6	26.6	0	1	"	-	-	-
N. Hexadecane	27.6	27.6	0	1	"	-	-	-
Toluene	28.5		0	1	"	-	-	-
P. Methyl-alkyl-siloxanne	28.9	26.9(2)	11.5	0.98±0.01	29	57.4	24.4	33
Dioxanne	33	33	0	1	γ_L	-	-	-
Di Propylene Glycol	35.5		0	1	"	-	-	-
D.M.F.	38	32.4	0	0.92-1	"	-	-	-
Pyridine	38	37.2	0	1	"	-	-	-
Propylene Glycol	38.4		22-37	0.8-0.93	32-35	71	56	15
Propylene Carbonate	40.5	36	0	1	>40.5			
Bromo Naphtalene	44.6	47±7(2)	22	0.93	41.5	86	87	0
DMSO	44.9	34.8(2)	25.4-33	0.84-0.9	39.5	84	76.7	7
Triethylene Glycol	47.4	29.5	33-37	0.78-0.84	40	86	70	16
Ethylene Glycol	49.7	41	45	0.7	39	-	-	-
Methylene Iodide	50.8	48.5±9	31.4	0.76-0.86	41.5	92	92	0
Formamide	58	39.5±7	44-50	0.65-0.7	41	97.5	81.5	16
Glycerine	65.7	37±4	57-65	0.48	42	97	79	18
NAOH 0.1N	72	22	50	0.64±0.04	52	118	61	57
Water	72	22	59-69	0.52-0.6	45	104	61	43
NaCl (Saturated)	82	22	68	0.36	40	112	61	51

Table 1. Spreading of solvents on an evaporated film of gold

(1) Values of Rhone Poulenc
(2) $\gamma_L^d = \gamma_L^{-2}$ (cf F.M. Fowkes in Ref. 7)
Values without references measured in this study.

2.6. Adhesives

The adhesive of reference is a mixture of the epoxide resin Shell 828 EL
(electronic grade) and the curing agent tri-ethylene tetramine in a ratio 100
to 8 and polymerised 2 hr at 60°C. Some room curing modified epoxies products
of Ciba Geigy have also been included in table V and VI, and the one component
epoxy adhesive AV130 with curing agent HY998 in table VII.

2.7. Joint Strength Testing

Adhesives joints were tested in a S.A.D.A.M.E.L.* micromachine Mi44 at a
speed of 5mm/min. Various geometries of the joints have been explored as shown
on fig. 6.

3. CHARACTERISATION OF THE GOLD SURFACE WITH A LIQUID CRYSTAL PROBE

The orientation of molecules in a thin film of liquid crystals may easily
be determined by either optical or electrical measurements.

The elongated molecules of liquid crystals may be parallel with or stand
perpendicular to the surface. Their orientation is related to the solid wall
surface of either low ($\gamma_s > 25$) or high ($\gamma_s > 100$) energy. (10)

The experimental value of the capacity of a thin layer of liquid crystals
made of 4 alkyle 4' cyanobiphenyles indicates that the molecules are in the
parallel orientation.

That is, a gold surface does not show any peculiarity toward liquid
crystal orientation, it compares to a polar polymeric surface.

4. WETTABILITY STUDIES

4.1. The Spreading of Various Liquids

In this paragraph only evaporated gold surfaces are considered. They
give very reproductible results. As we considered different aspects of the
wettability of the gold surface, the contact angle θ of 27 different liquids, as
indicated in table 1, were measured. From the experimental values it is clear
that liquids of surface energy below 40m JM^{-2} spread over the gold surface and
that the critical tension is between 40 JM^{-2} and 45m JM^{-2}. Only a polymethylal-
kylsiloxanne (R.P.V 500/308) and propylene glycol give different results.

* SADAMEL 2300 La Chaux-de-Fonds, Switzerland.

4.1.1. The Adsorption of Polymethylalkoylsiloxanne.

In order to measure the surface critical value, liquids of various surface tension had to be chosen. Among the liquids available was a polysiloxanne oil of $\gamma_L = 29m$ JM^{-2}. This oil is described in the commercial brochure* as a methyl alkyl polysiloxane (V500/308) formula

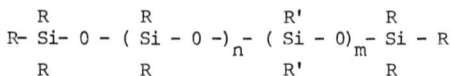

$$R- \overset{\displaystyle R}{\underset{\displaystyle R}{Si}} - 0 - (\overset{\displaystyle R}{\underset{\displaystyle R}{Si}} - 0 -)_n - (\overset{\displaystyle R'}{\underset{\displaystyle R'}{Si}} - 0)_m - \overset{\displaystyle R}{\underset{\displaystyle R}{Si}} - R$$

Where R is a methyl group and R' is not precisely specified.

As we were drawing Zisman plots for various different substrates it became became apparent that independent of the substrate, the V500 silicone oil always gave the same contact angle. This is also true for the gold surface indicating a strong adsorption of the oil which leads to an apparent surface tension $\gamma = 29$, JM^{-2}.

This behaviour recalls that of autophobic films and is not observed with other polysiloxane   It disappears (the oil wets) when the gold surface has been treated with a silane although the silanation is not otherwise visible (see later). As it may be relevant to the lubricating properties of this oil it was thought to be worth reporting.

For comparison, toluene, which has similar (28.5 JM^{-2}) surface tension value, entirely wets the gold surface.

4.2. Zisman Plot

If one measures the contact angle, θ, of a series of hydrocarbons of different surface tension, γ_L, on a given surface, a plot of cos θ versus γ_L will show a linear variation. The extrapolation of that line to the value cos θ =1 defines the critical value of γ_c, which is the value of a liquid that just wets the surface (11). When γ_c is higher than the surface tension of a hydrocarb (i.e. 35m JM^{-2}), polar liquids have to be used. Then instead of a straight line, a band of cos θ (γ_L) values is defined, which gives a range for the solid critical tension (fig.3). In our case 33 < γ_c < 40m JM^{-2} with a mean value of $\gamma_c = 37m$ JM^{-2}. As previously mentioned the silicone oil V500 and propylene glycol adsorb and stay out of the range. The results of the wettability studies described in paragraph 4.1 and 4.2 indicate that gold has a low surface tension.

* Rhone-Poulenc Rhodorsil oils.

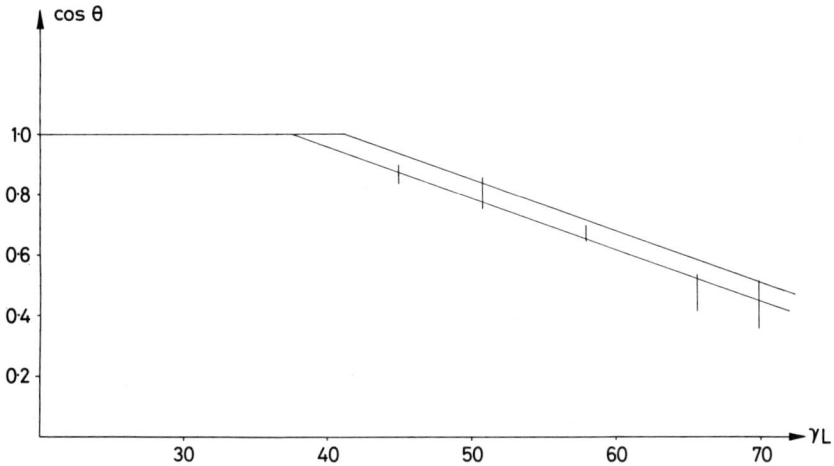

cos θ

Fig. 3.   Zisman's plot of the contact angle of various organic
               liquids on gold versus their surface tension.

4.3.   The Dispersive Part of Solid Gold Surface Tension

Interactions of liquids with surface may be described (12) as the sum of
their dispersive (γ_L^d) and polar (γ_L^d) interactions. Measuring the contact angle
of a liquid having only dispersive interactions with a surface allows the
dispersive contribution to the surface tension to be obtained.   Two such
reference liquids, methylene iodine $(\gamma_L = 50.8)$ and αabromo-naphthalene $(\gamma_L = 44.6)$
are widely used (7). Their contact angles on gold lead to the same value of the
dispersive energy of interaction (Table II) $\gamma_L^d = 42 \pm 2$. Then, as is often
the case, the solid critical value is nearly equal to the surface tension due
to dispersive interactions. From these results it would appear that gold is a
purely dispersive surface of low surface tension.

In fact one may suspect that some contaminants adsorb on the gold surface
and that the previous methods are unable to reveal its presence.

Test Liquid	θ	γ_S^d
Methylene Iodide $\gamma = 50.8$	31 — 40	42 ± 2
Bromo Naphthalene $\gamma = 44.6$	22	41.5 ± 1

TABLE II   The dispersive interaction of gold with non-polar organic liquids

There have been new developments in wettability studies which suggest that the bare solid surface tension may be attained independently of liquid adsorption. These are considered next.

4.4. Neuman's Equation of State

Neuman et al (8) have devised an equation of state which relates contact angle, liquid surface tension and solid surface tension. This very convenient expression has been tabulated (8b) allowing one to obtain γ_S knowing θ and γ_L. Adsorption phenomenum seem to be unimportant in this treatment. Values are reported in the "γ_S" column of Table I. They are found in the range $\gamma_S = 41 \pm 3$ for the non wetting liquids.

Again the value obtained is in accordance with the previously indicated results. This is a fairly low value for a metal, Silver, for example has a γ_L^d value of 70 and Copper a value of 60 (7) Neuman's equation probably relates to the apparent surface tension; a layer adsorbed from the atmosphere lowering the bare metal surface tension γ_o to an apparent value of the surface $\gamma_S = \gamma_o - \Pi e$. Liquid adsorption has been considered by Chappuis (9) and we also looked at the application of his treatment to our problem.

4.5. Chappuis's Theory

Recognising that a liquid will adsorb on a solid of higher surface energy and lower its surface tension, the relation between contact angle and surface tension is written:

$$\cos \theta = \frac{2\gamma_o - \gamma_L - \Pi e}{\gamma_L} \qquad \ldots\ldots (2)$$

From the definition of the work of adhesion Wa (Wa $= \gamma_L(1 + \cos\theta)$) and the energy of cohesion of a liquid Wc (Wc $= 2 \gamma_L$). Equation (1) may be expressed as:

$$Wa - Wc = -2 (\gamma_L - \gamma_o) - \Pi e \qquad \dots (3)$$

where Wa - Wc is the spreading coefficient S. The lower the liquid surface tension the higher should the spreading pressure be. The graphic representation of Wa - Wc for different liquids as a function of their surface tension will reveal any change in the value of the adsorbed layer spreading pressure. Liquids of high surface energy cannot adsorb ($\Pi e = 0$) and will give a straight line which will cut the horizontal axis when γ_L is equal to the surface tension γ_o of the bare solid.

The plot resulting of our results on gold is shown in fig. 4. The four liquids of higher surface tension (methylene iodide, formamide, glycerol and water) fall on straight line of slope -2 as expected on a bare surface for which the γ_o value is 45. All the different mathematical treatments of the wettability concur in characterising gold as having a relatively low ($42m$ JM^{-2}) surface energy of mainly dispersive character. Nevertheless, the polar liquids of high surface energy show a contact angle corresponding to both dispersive and polar interactions.

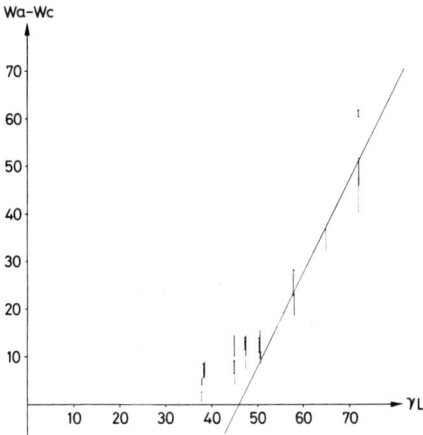

Fig. 4.   Chappuis's plot of the spreading coefficient Wa-Wc versus contact angle, —— gold plated ⋯⋯ pure gold evaporated —x—x—. massive gold.

4.6. The Surface Energy of Gold due to Polar Interactions

The energy of adhesion, Wa, of a liquid to a solid may be split into contribution due to dispersive interaction Wa^d and those due to polar interactions, Wa^P

$$Wa = Wa^d + Wa^P \qquad \qquad \text{.... (4)}$$

Wa is calculated from the liquid contact angle as

$$Wa = \gamma_L (1 + \cos\theta) \qquad \qquad \text{.... (5)}$$

and Wa^d is the geometric mean of γ_s^d and γ_L^d

$$Wa^d = 2\sqrt{\gamma_s^d \, \gamma_L^d} \qquad \qquad \text{.... (6)}$$

If, as our results tend to indicate, gold was a purely dispersive surface then one would get for water an energy of adhesion $Wa = Wa^d = 2\sqrt{42 \times 22} = 60m\ JM^{-2}$ instead of the value of 95m JM^{-2} obtained from equation (5). Inserting these results in (4) the value of the polar energy $Wa^P = 76m\ JM^{-2}$, of interaction of water with gold comes out. Using the value $\gamma_L^d = 42m\ JM^{-2}$, Wa^P may be calculated for the various polar solvents of high energy. Results are collected in the last column of Table I and show a non-negligible polar interaction energy. Using the formalism

$$Wa^P = 2\sqrt{\gamma_s^P \, \gamma_L^P}$$

which may be valid in a limited range of γ_L^P values. One finds a polar contribution $(\gamma_s^P \ 3 - 12m\ JM^{-2})$ to the gold surface tension which depends strongly upon the acidity of the solvent.

4.7. The Acid Base Character of the Surface of Gold

The origin of the polar interactions are presently thought to be essen-- tially acido-basic in character (13). The contact angle values of Table I definitely show deviation in the case of acidic glycols or basic solvents.

Drops of some typical acidic and basic liquids have been applied to a gold surface and their energy of adhesion compared to the energy of dispersive interactions, admitting that the gold surface energy is unknown (γ_s^d). Results are collected in Table III. Both acids and bases interact with the gold surface; stronger bases have the higher adhesion energies. Acids interact more than other liquids showing the strong basic character of the surface. Although the contact angle of water shows wide variation in different measurements, a 0.1N solution of NaOH has very reproducible value. A 0.1N HCl solution spreads over gold in a few seconds indicating that if contaminants are adsorbed onto gold they are displaced by an acid.

Wettability studies show that the gold surface under ambient conditions has a moderate surface energy of $42m$ JM^{-2} with mainly dispersive interactions. Acidic or basic solvents reveal polar interactions. The uncertainty on γ_{au}^{o} precludes a quantitative characterisation of the acid-base character of gold.

Liquid	γ_L	γ_L^d	cos θ	Wa	Wa^d x γ_s if γ_s^d=42		Wa^p (γ_s^d = 42)
Acid							
HCl 0.1N	72	22	0.96±0.04	>140	9.4	61	>80
Glycerol	65.7	37	0.48	97.5	6	39.4	58
Water	72	22	0.39±0.04	100±3	9.4	61	42
p.methyl alkyl⁻ siloxanne	29	27	0.97	57	10.4	67.4	0
DMF	38	32	0.97	75±2	11.3	73	0.2
DMSO	45	35	0.88	84	11.8	76.7	7
T.E.T.A.	39.7	39.7	0.87	74.5±1	12.6	81	0
Pyridine	38	37	1	>76	12	77	≥1
NaOH 0.1N	72	22	0.64	118	9.4	61	47
Base							

TABLE III Acid-base Interaction between Gold and Organic Liquids.

5. EXTENSION TO OTHER GOLD SUBSTRATES

The previous results so far reported concern an evaporated film of gold. In general gold is not used in the form of a pure metal but either lightly alloyed or galvanically plated. We have compared the wettability of pure flat molded gold, alumina polished 18ct Gold-silver, 18 ct. Gold-Cadmium plating, 22ct Gold-silver plating with that of evaporated layers.

If the results show a 10% variation and a great variability in the contact angle of water, the surface tensions is always found in the 37-45m JM^{-2} range. Alumina polishing does not produce water spreading. The values reported in Table IV indicate that the preceeding results are valid for these various surfaces also.

Liquid	Evap. Gold	Gold Metal (18 ct)	Gold Plated (22 ct)
Water	0.36 - 0.52	0.43 - 0.75	0.51 - 0.68
NAOH	−	−·	0.61 - 0.68
Glycerol	0.43 - 0.54	0.59 - 0.65	0.21 - 0.37
Formamide	0.65 - 0.70	0.52 - 0.6	0.73 - 0.81
CH_2I_2	0.76 - 0.86	0.6 - 0.7	0.75.5
T.E,G,	0.78 - 0.84	0.7 - 0.87	0.69 - 0.76
P.G.	0.86 - 0.93	0.78 - 0.85	0.83 - 0.91
D.M.F.	0.94 - 0.98	1	0.89

TABLE IV   Comparison of Wettability of Various Gold Surfaces −
Values of cos θ

This is further evidenced by the graph of Wa − Wc versus γ_L for the various substrates (fig.4).

The difference between a freshly cleaned surface either from melt (3) (4) or U.H.V. (2) and a surface exposed to atmosphere is not known.

Gold plates cleaned in a detergent solution are wetted by water. As soon as they are dried either with a hot air gun, spin drying or from alcohol they turn hydrophobic. The usual explanation by atmospheric contamination is an easy one which is not well documented.

Zisman's (6) observation that the same surface tension values was measured on various inorganic surfaces and that the γ_C values depended upon the atmospheric humidity led him to the acceptable suggestion that a water layer was adsorbed on the surface. Effectively, when we tried to obtain gold γ_S^d value from two liquid surface tension measurements (14) the situation was that water was not displaced by the organic solvents and remained on gold as a spherical drop in any organic liquid. Under water, all organic solvents fail to wet the surface. From displacement criteria (15) this shows that the polar interaction of gold and water are very high. Nevertheless the presence of a water layer cannot be accounted for by a spreading pressure value πe satisfying both relations $\gamma_s = \gamma_s^o - \pi$e and πe $= W_a^d - 2\gamma_L$, and the value of the surface energy does not correspond to that of water.

This subject is not well understood but clearly, the practical gold surface has a surface tension of 42m JM^{-2} and interacts with most organic compounds through dispersive forces.

6. THE BEHAVIOUR OF EPOXY ADHESIVES

The consequences of the preceeding results is that, in the course of deposition of an adhesive on a gold surface the adhesion will encounter a low energy surface. The behaviour of some room curing epoxy adhesives during the polymerisation was observed and it will now be described.

6.1. Epoxy Resins

The surface tension of epoxy resins may be measured by the Wilhemy method (or the hanging drop for highly viscous resin). Using the former we found, at 20^oC, values in the region of 48m JM^{-2} for various DGEPA and DGEBA modified resins. The value of 49 ± 0.2 for Epon 828 EL compares well with the 46.2 value obtained at 25^oC by the hanging drop method (16). Our results are collected in Table V.

When the resin is deposited over a gold surface it shows the contact angle value expected for a surface of 42m JM^{-2}.

Resin	Y_L	n	$Cos\theta$	Curing Agent	Y_L	n	$Cos\theta$	Parts curing agent per 100 parts resin in mixture	Y_s	Y_s^d
DGEBA	49.2	120	0.62–0.71	Polyamine	39.7	10	0.85–0.91	8	39	39
DGEBA	48.7	120	0.57–0.69	Polyoxyethy-lene amine	32.1	9	0.95–0.98	30	41	43
DGEBA base flexibilized	39.4	300	0.91–0.99	Polyamine Modified	43.4	450	0.83–0.86	23	57	40
DGEBA base modified	44.1	5000	0.76–0.97							

TABLE V   Surface Tension of Epoxy Resin Curing Agents and Adhesive Mixtures

6.2.   Curing Agents

Room temperature curing is obtained by adding amines to the epoxy resin. Amines have low surface tensions (table V) of the order of 30m JM^{-2}. Because of this low surface energy they spread on the gold surface.

Both the resins and the curing agent wet gold as predicted by Dupre equation.

6.3.   Mixture of Resin and Curing Agent

The mixture of the resin (Shell 828 E.L.) and curing agent T.E.T.A. (triethylene tretramine 8%) behaves as a solution of surface tension which is a weighted average of that of its components. The polymerised resin has the same surface energy as the mixture which demonstrates that the surface tension does not change during the polymerisation reaction. Mixtures of surface tension higher than 42m JM^{-2} do not wet the gold initially but as the polymerisation proceeds it tends to spread. As the adhesive mixture surface tension does not change, this shows that the apparent surface energy does and recalls the behaviour of an acid solution. The change of contact angle with time is shown in Fig. 5. All the resins considered and mentioned in table V, show the same process. Often after gelification the contact angle variation slows down and polymer retraction may be measured from the final dimension of the drop.

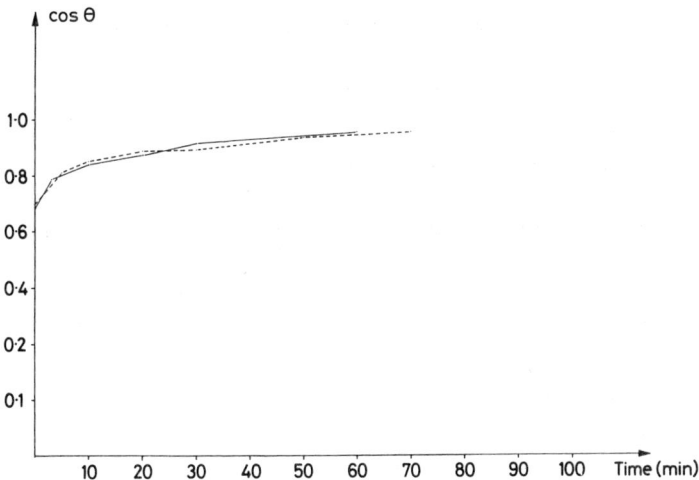

Fig. 5.   Variation of the adhesive contact angle with time.

7.   SILYLATION OF GOLD

Treatment of surfaces with a silane primer is widely used as a way of improving adhesive adherence. Although the mechanism is not demonstrated it is thought that silanol groups produced by the hydrolysis of the silane react by a condensation reaction with oxides on the surface. Gold does not form any oxides at room temperature (17) (18) and this mechanism cannot hold. Silanation of surface could improve adhesion through the formation of polysiloxanes which adsorb strongly.

The priming of the surface with four compounds: VIMO (vinyl-trimethoxy silane); GLYMO (glycidyloxy propyl trimethoxy silane); AMEO (aminopropyl tri-methoxysilane) and MEMO (methacryloxy trimethoxysilane), was considered.

Wetting of the surface demonstrated a slight increase in surface energy which became $45m \ JM^{-2}$. Also the polymethyl alkylsiloxane V500-308 which did not wet the clean surface (see para.2) now spread over the silylated gold showing a definite influence of the silanation. But the joint strength value did not change for any of the primers considered, with the exception of a narrower variation of the experimental values around the upper values. Although their efficiency is not as apparent as it is in case of glass or steel, silane may improve joint strength in the weakest cases.

8.   JOINT STRENGTH

Our application concerns small surfaces and we have to scale down the usual normalised conditions, particularly when working with costly gold specimens. Our tests used one of the six configurations sketched in fig.6.

8.1.   Lap Shear Tests

The joints were made between two rectangular plates of gold-plated brass which overlap each other by $10 \ mm^2$. The joint thickness was of the order of $0.1 \ (\pm \ 0.02)$mm. The rupture of the joint was adhesive with room curing epoxies. The results shown in the first column of table VI are difficult to compare with the literature data which relate to aluminium. Their meaning in comparison with other materials will be discussed in paragraph 9.6. The initial shear strength is good.

8.2.   Tensile Strength

The dimension of our samples are given in fig. 6B and the results in table VI column B, when available. The values are obtained at the upper limit of the apparatus and the comparison with the blister test that should give the

187

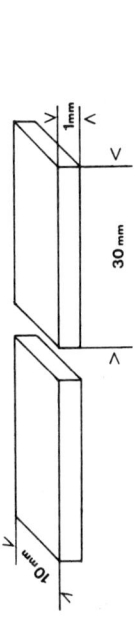

6. A

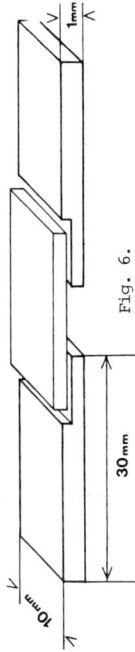

6.B

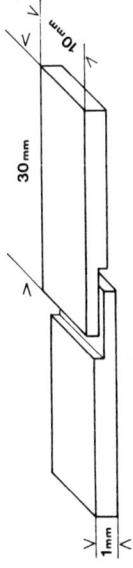

6.C

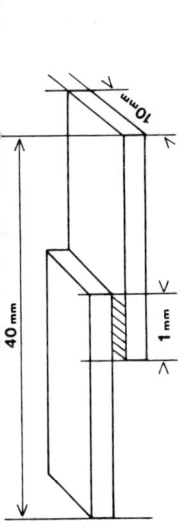

6.D

Fig. 6.

Geometry of the specimens used to characterise bonding to gold.

A   Shear test
B   Tensile test
C   Step joint
D   Double step joint
E   Blister test
F   Wedge test

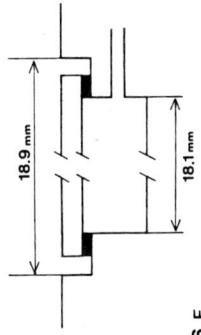

6.E

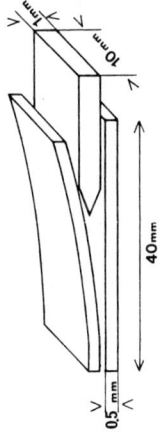

6.F

ADHESIVE	A Shear Test	B Tensile Strength	C Step Joint	D (2) Blister Test	E Wedge Test
1. Modified DGEBA	42 ± 8	–	20 ± 4	33	19
2. Flexibilised DGEBA	15 ± 3	–	14 ± 4*	11	30
3. Mixture of 1 and 2 25.75	42 ± 3	–	23 ± 5	55	14
4. idem 50.50	34 ± 3	–	15 ± 5	71.5	20
5. 823 + TETA	37 ± 7	40	45 ± 5	33	16
6. DGEBA + polyoxyethylene amine	61 ± 7	–	22 ± 2	55	19
7. AY139 + HY909	40 ± 2	40 ± 5	–	66	10
8. one component DGEBA	> 55 ( )	> 50 ( )	–	100± 10	15
9. one component modified	52 ± 10	> 50 ( )	–	155	14

TABLE VI   Strength of Joints to Gold with Various Adhesives   (all in MPa except E where L_o is mm of fracture)

same value is not feasible.

8.3. "Stepped" Joints

In order to increase the joint resistance we considered "step" joints following the design of fig. 6C and 6D. As the tensile strength is in general lower than the shear strength, and that we have related the measured force to the entire surface of the joint the stepped joints, resistance appears smaller than the shear resistance although it is not if one considers only that surface of joints which is stressed in shear.

Adhesion to glass is higher than to gold and both design 6C and 6D give the same results, as the joint breaks at the gold surface.

8.4. Blister tests

The "blister test" (19) has been adapted in the following way: a step of about 0.5 x 0.5mm is formed at the upper end of a cylindrical box (diameter d) a disc of glass is stuck to it (fig.6E). The surface of the joint is

$$S = \frac{\pi}{4} \ (d^2 - d'^2) = 0.23 cm^2.$$

A pressure P is applied by a compressed gas injected at the side through a small tube and the pressure at which the assembly breaks is noted. The applied force is

$$F = P_x \frac{\pi}{4} d'^2$$

Then the tensile resistance by surface unit is $F/s = pd'^2/d^2 - d'^2$. The results of the measurement should compare with the tensile strength. In fact they are slightly higher (60 instead of 40 MpA) probably due to a better definition of the joint.

8.5. Wedge Tests

The wedge test (20) is very convenient to evaluate the environmental resistance of an adhesive (see par.9). We used a small rectangular sheet (10x40x0.5mm) of hard stainless steel (17-8) AISI 301 which have been gold plated with 5µ gold 18 ct. and convered with a flash of 22 ct. gold. They are stuck together with an adhesive the thickness of which is controlled by two spacers of Kapton (0.075mm). After curing the thickness is measured with a wire gauge which is kept at the sample width. The Kapton spacer is taken out and a wedge of 1mm thick is pushed in. The adhesive joint cracks over a distance L_o, L_o is measured with the gauge mentioned above as that distance of the wedge where the opening does not pass through. L_o is indicative of the adhesive peel strength (21). The values measured with different adhesives are collected in column E of table VI.

8.6. Comparison with Other Substrates

As our test specimens have different sizes from the normalised test requirements, comparison with literature data is not easy. We have thus compared tensile and shear strength of the same adhesive (joining different metals, i.e. brass, steel and stainless steel. The adhesive retained may not have been a best choice as it never gives a purely cohesive fracture. So results obtained with a one component high temperature epoxy resin (AV 118) have been included. Nevertheless results, as shown in table VII, indicate that the initial strength is as good with gold as with other metals.

ADHESIVE   828 /TETA

Substrate	Tensile strength	Shear strength	Peel
Brass	35±5	40±5	-
Stainless steel	35±5	38±6	20
Steel	36±9	38±6	-
Gold	42±8	37±5	16

ADHESIVE 138/998

Substrate	Tensile strength	Shear strength
Brass	40±3	46±4
Stainless steel	44±4	45±1
Steel	40±8	47±3
Gold	35±8	40±2

TABLE VII   Comparison of the joint strength between different metals and gold.

9. ENVIRONMENTAL RESISTANCE

One critical property of adhesive joints is their resistance to the environment mainly water. As gold does not corrode one may think that joint durability will be higher. This is true as shown by the comparison of joint strength of adhesive joints between brass and gold after exposure in humid atmosphere. But comparison of gold with stainless steel evaluated with the wedge test shows an extreme sensitivity of the gold joint - table VIII. Adhesive joints to gold are also very sensitive to dimethylformamide.

Note that in table VIII the one component epoxy resin does not show any sensitivity to moisture if used with stainless steel but weakens when joining gold substrates. In every case the rupture of the humid joint is adhesive.

Metal Adhesive	Gold			Stainless Steel		
	L_o (mm)	L_{24} (mm)	ΔL (mm)	L_o (mm)	L_{24} (mm)	ΔL (mm)
828 / T.E.T.A.	16	26.5	10.5	16.5	20.5	4
Modified DGEBA One component	14	20.5	6.5	11.5	14.5	3
Epoxide resin One component	15	19.5	4.5	10	10	0
Flexibilized Epoxy	20.5	37	16.5	20	30.5	10.5

Table VIII   Results of Wedge crack test in humid environment (40°, 93% R H) Comparison of the stability of joints made between stainless steel or gold and exposed to a tropical atmosphere.

The extension of the initial crack is indicative of the sensitivity of the joint to mixture.

(1) L_o initial length of the crack after introduction of the wedge

(2) L_{24} length of the crack after 24 hr at $40^\circ C$ and 93% Relative Humidity

(3) $\Delta L = L_{24} - L_o$

192

10. CONCLUSIONS

Gold has an apparent low surface energy ($42m\ JM^{-2}$). The initial tensile strength of adhesive joints is higher than would be predicted from that surface tension value (22). This results from the change of the gold surface energy during the adhesion reaction. Joint strength is improved by an aminosilane primer. Although the initial strength of adhesive joints with gold is as good as with stainless steel it is very sensitive to humidity and solvent vapours.

ACKNOWLEDGEMENTS

We thank M.M.F. Chabloz and C. Ganguillet for their measurements of the joint strength and K.W. Allen for editing the text.

Part of this study has been financed by the National Swiss Impuls Programme.

REFERENCES

1.   J. Schultz and A.N. Gent, J. Chim. Ohys. 70.708 (1973)
2.   T. Smith J. Coll. Int. Sc. 75 51 (1980)
3.   G.L. Gaines ibid. 79. 295 (1981)
4.   M. Schneegans and G. Menzel ibid. 88. 97 (1982)
5.   K.W. Bewig and W.A. Zisman J. Phys. Chem. 69.4238 (1965)
6.   M.K. Bernett and W.A. Zisman J. Coll. Int. Sc. 28 243 (1968)
7.   F.M. Fowkes "Surface Chemistry" in Treatise on adhesion and Adhesives R.L. Patrick Edit. vol. 1 M. Dekker (1966)
8a.   A.W. Neumann, R.J. Good, C.J. Hope and M. Sejpal, J. Coll. Int. Sc. 49 291 (1974)
8b   Tabulated values are given in: Separation and purifications methods 9(1)69-163 (1980)
9.   J. Chappuis in :Multiphase Science and Technology G.F. Hewitt, J.M. Delhaye and N. Zuber Edit. Hemisphers Publ. Corp. (1982)
10.   J.J. Cognard Mol. Cryst. Liq. Cryst. Suppl. 1 (1982)
11.   M. Zisman Adv. Chem. Ser. 43.1 (1964)
12.   F.M. Fowkes ibid. p. 99

13.   F.M. Fowkes in Microscopic Aspects of Adhesion and Lubrication J.M.Georges
      ed. p.119-139 Elsevier 1982.

14.   J. Schultz, K. Tsutsumi and J.B. Donnet, J. Coll. Int. Sc. 59.272 (1977)

15.   M.E. Shanahan, C. Cazeneuve, A. Carre et J. Schultz J. Chem. Phys.
      79.241 (1982)

16.   H. Dannenberg and C.A. May Treatise on adhesion and adhesive vol. II p.18
      R. Patrick Ed. M. Dekker (1967)

17.   An electrochemical, intensity versus potential, plot on a gold electrode
      heated to 800°C in air does not show any cathodic reduction wave not anodic
      oxidation as opposed to other metals, platinum included.   J. Clavillier
      private communication

18a. D. Clark, T. Dickinson and W.N. Mair J. Phys. Chem. 65. 1470 (1961)

18b. D. Clark, T. Dickinson and W.N. Mair, Trans. Far. Soc. 55. 1937 (1959).

19.   T.V. Parry and A.S. Wronski Adhesion 5, K.W. Allen Edit. Chap. 1 (1981)

20.   G.P. Anderson, S.J. Bennett and K.K. de Vries, Analysis and testing of
      adhesive bonds Acad. Press, London (1977).

21.   G.P. Anderson, (Thiokol Corp.) private communication.

22.   W. Wake "Adhesion and the formulation of adhesives" p. 58, Appl. Sc. Publ.
      (1976)

Chapter 11

ADHESION PROBLEMS IN PLYWOOD MANUFACTURED FROM TROPICAL HARDWOOD SPECIES

J D WAGNER and D R BAIN

Unilever Research, Port Sunlight Laboratory

INTRODUCTION

The tropical rain forest, from which UAC (Nigeria) Ltd receives hardwood logs for plywood manufacture, comprises a very large number of plant species. About 20 of the tree species, found within economic range of the mill at Sapele, provide logs suitable for plywood manufacture.

The plymill was established in 1947 using phenol formaldehyde (PF) based adhesives for making weather and boil proof plywood to BS1455, and for marine grade to BS1088.

In 1964 the local manufacture of PF resins was started. Considerable operating experience has now been acquired in the plymill, using various glues based on our own PF resins. The number of wood species that are feasible for plywood manufacture has meanwhile been increased to include several that were originally too problematical.

The standard glue formulation is satisfactory for the bulk of this group of hardwoods. However, it has long been known that certain species are particularly difficult to glue unless the adhesive and/or the production conditions are modified.

This paper defines the phenomenon of "difficult-to-glue", proposes some reasons for its occurrence and reports various experimental measures that have been taken to secure satisfactory plywood bond quality - again as evaluated per BS1455.

1 THE PLYWOOD MANUFACTURING PROCESS

Plywood is a multi-ply board product in which layers of wood veneer and thermosetting adhesives are consolidated in a hot pressing operation. Veneers for plywood manufacture can either be sawn from a squared log using a veneer saw, sliced from a quartered log (referred to as sliced veneer) or peeled in a continuous sheet from a rotating log (rotary-peeled veneers).

As it is peeled, the veneer is re-wound onto a reel in such a way that the artificial inner and outer veneer surfaces are reversed. This deformation of the sheet demands stress relief of the veneer in order to achieve satisfactory and rapid reeling. It is possible because of the presence of thousands of fissures extending part way through the veneer parallel with the knife edge of the peeling lathe. These "lathe-cheeks", as the fissures are known, arise on the centripetal surface of the veneer as it emerges from a zone of pre-compression induced in the immediate vicinity of the knife edge.

The veneer is then dried to 6-8% MC in a continuous sheet before cutting to size and assembling into plyboards as indicated in Fig. I.

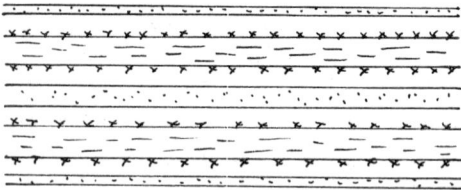

Fig. I   Typical lay-up of a 5-ply board

The glue for this experimental programme was made up of three components mixed in the following proportions:

Resin	100 parts by weight
Extender	20 parts by weight
Filler	10 parts by weight

This glue was spread onto the crossband veneers at the rate of
170 g/m^2. The boards laid up as indicated in Fig. I were cold pressed
at 5 kg/cm^2 for 5 min. This consolidated the board by causing glue
transference from crossband to adjacent veneers (which also makes hot
press-loading much easier on full scale working).

The cold pressed boards were left for a further dwell period
(referred to as the 'closed assembly time') then hot pressed at 127°C and
10-12 kg/cm^2 for a time determined by the board thickness.

After hot pressing the boards were 'hot stacked' for 24 hr before
trimming and bond evaluation.

The bond quality in our plywood is assessed by the knife test
described in BS1455:1972. This involves cutting 5 " square samples from
the board, breaking the bond with a special knifing tool and assessing the
exposed surfaces for wood failure. Numbers, on a scale, from 0-10
(illustrated in BS1455) are given, 10 being equivalent to 100% wood
failure i.e. an excellent bond. For the board to be classed as WBP (water
and boil-proof) it is steamed for 16 hr under pressure at 120°C, cooled
under water and knifed again. For our experimental boards the 24 hr cold
water soak test laid down in BS1455 for WBP standard was omitted to
simplify the test results. In both cases an average value of 5 must be
obtained with no individual value lower than 2.

2 DIFFICULT-TO-GLUE SPECIES

As pointed out in the Introduction, about 20 different species of
timber are used in plywood production. Table 1 gives a list of the most
common species and the typical proportions in which they arise.

Table I

Hardwood species used for plywood manufacture in Sapele

Timber Species		% Total Mill Input Volume
White Afara	(Terminalia superba)	20
Lagoswood	(Khaya ivorensis)	20
Obeche	(Triplochiton scleroxylon)	20
Pterygota	(Pterygota bequaertii)	6
Sapelewood	(Entandrophragma cylindricum)	6
Antiaris	(Antiaris africana)	5
Daniellia	(Daniellia ogea)	5
Canarium	(Canarium euphyllum)	5
White Sterculia	(Sterculia oblonga)	4
Other		9
		100

Normally these species are mixed in preparing standard plywood boards and most species form excellent bonds to other species particularly Obeche and Antiaris. However, it has been known for many years that certain species when self-bonded give good bonds only under modified plymill conditions using type "A" glue and this is demonstrated by the results in Table II.

Table II

Dry knife values for single species plywood boards

Species	Dry Knife Values	Average Dry Knife Value
Lagoswood	5.4.3.3.4.4.	3.8
Afara	6.4.2.2.4.6.	4
Sterculia	4.2.2.0.0.4.	2
Celtis	7.4.3.2.3.7.	4.3
Daniellia	6.5.3.3.2.6.	4.2
Antiaris	7.8.6.8.8.7.	7.3
Obeche	8.5.6.7.5.7.	6.3

In the few instances when 'difficult species' were used e.g. marine grade, which is prepared entirely from durable redwood species, satisfactory adhesion was obtained by 'double spreading' i.e. gluing both sides of all the internal veneers, and one side of the outermost veneers.

Possible mechanisms put forward in the literature for the bonding of wood are:-

(a) Mechanical interlocking of cured glue anchors in the wood cellular structure.

(b) Hydrogen bonding of polar groups in the resin molecules with active sites in the wood structure.

(c) Covalent bonding of the resin to the wood.

It is now thought that purely mechanical interlocking, ref. 1, i.e. 'lock and key' bonding plays very little part in forming adhesive bonds and that hydrogen bonding, refs. 2,3, is the predominant mechanism. In some cases e.g. urea glues, reaction with the wood cellulose has been shown, ref. 4, to give some covalent bond contribution. Recent work by Klein, ref. 5, who examined glue-lines by microscopy suggested an optimum bond strength was obtained by glue penetrating 8 cells deep each side of the glue line. Hence to form a good bond the resin must:-

(i)   Have sufficient reactive groups for hydrogen bonding to occur.

(ii)   Be able to wet and penetrate the wood.

(iii) Have molecules sufficiently mobile to align with reactive sites in the wood.

(iv)   Cure with the minimum of shrinkage to preserve hydrogen bonds.

To show that wetting and penetration of glue into different timbers mirrored the quality of the bond, 7-ply plywood boards were prepared in the laboratory using the above-mentioned pressing conditions from all

Obeche, Afara, Lagoswood and Sterculia veneers, and a strip was cut from
the boards at an angle of 45° to enable the glue lines to be examined.
Photographs of the bonded glue lines are compared in Fig. II (i-iv)
which clearly shows a marked difference in glue diffusion into the veneer
for different species Lagoswood and Afara indicating high glue penetration
whereas virtually no diffusion into Sterculia or Obeche is occurring.

From this qualitative assessment it would appear that Lagoswood and
Afara behave in a similar fashion i.e. glue is diffusing away from the
glue line possibly leading to starved glue lines which would certainly
give a poor bond. The diffusion into Obeche and Sterculia, however,
appears to be similar whereas their gluing ability is markedly different.

A resol-type phenolic resin, prepared by reacting phenol and
formaldehyde in alkaline solution consists of a mixture of polymers
containing the following basic repeating units.

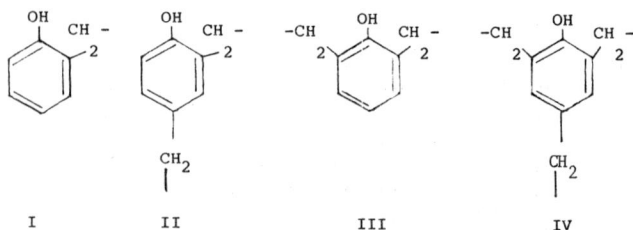

Spectroscopic analysis of freeze dried 'A' type resin samples
indicates that all the polymer linkages are methylene i.e. $-CH_2-$ groups
and total acetylation indicates the polymer to be in a cross linked
configuration i.e. predominantly units I, II and IV linked together
rather than in linear conformation i.e. totally chains of units I and III.

Although this analysis confirms the chemical nature of phenolic
resin it does not indicate the size or molecular weight of the polymers of
which the resin is made up. To determine the molecular weight
distribution (MWD) a technique called high performance gel permeation

FIG. II (i)

FIG. II (ii)

202

FIG. II (iii)

FIG. II (iv)

chromatography (HPGPC) is used, in which molecules are separated on a size basis as they pass through columns of crosslinked polystyrene beads of controlled porosity. Large molecules having access to the lowest pore volume elute first, followed by molecules of decreasing size.

Much of the work in the literature concerning phenolic resins refers to GPC of novolaks or acid catalysed resins which are readily soluble in tetrahydrofuran (THF) or dimethyl formamide, refs. 6,7,8, and therefore presenting little or no problem in analysis.

Phenol formaldehyde plywood resins however contain 10-20% sodium hydroxide and the problem in obtaining MWD data is one of rendering the material soluble in a suitable GPC solvent, preferably THF so that the sample can be run routinely on conventional GPC systems. The method described by Waters Associates, ref. 9, using acetic acid to neutralise and solubilise the resin was found to give incomplete dissolution of our samples and even the soluble part was very unstable and liable to precipitate from solution and Matsuzaki et al have commented on the insolubility of PF resins in THF in their work with DMF as solvent, ref. 10. Pizzi and Scharfetter described the use of trichloroacetic acid to neutralise tannin-formaldehyde resins for thin layer chromatography, ref. 11. This technique was applied to phenolformaldehyde resins and the use of a 10% solution of trichloracetic acid was found to effectively solubilise phenol formaldehyde resins in THF.

Samples were run on a modular HPGPC system consisting of a Waters M6000 pump, Waters U6K injector and a Cecil UV Spectrophotometer operating at 283 nm. A low molecular weight Styragel column set (100, 500, 1000°A) was used. The addition of a second 100°A column gave no significant increase in resolution of the low molecular weight components. To maximise resolution it is desirable to work with as few columns as possible, as long as the porosity is sufficient to span the entire molecular weight distribution of the sample. Solvent was HPLC grade tetrahydrofuran (Fisons or Rathburn Chemicals) which is free of UV absorbing materials.

Fig. III shows the traces obtained for our resin "A" and a
development resin "B", prepared by varying the formaldehyde:caustic ratio
and Dynosol ex Dyno Industries Ltd from which it is immediately obvious
that "A" contains significantly more low molecular weight material than
"B" and Dynosol as indicated by the large number of peaks on the right
hand end of the trace. From the relationship of diffusion coefficient D_0,
with molecular weight M,

$$D_0 = \frac{kt}{KfM^\alpha}$$

where k is Boltzmann constant, t = temperature and k,f and α are
constants, the low molecular weight components of "A" will diffuse more
rapidly from the glue line and not be effective in the adhesion.

Table III shows the average WBP knife values obtained from a number
of Afara and Lagoswood 7-ply 18 mm boards prepared in the laboratory using
resins 'A' and 'B' under our standard experimental conditions together
with two commercial glues Dynosol (ex Dyno Industries) and Aerophen (ex
Ciba Geigy) used according to the manufacturers recommendations.

Table III

Effect of resin on bonding

Resin	Average WPB Knife Values	
	18 mm Lagoswood	18 mm Afara
Aerophen	4.3	4.3
Dynosol	3.5	4.8
A	3.8	4.0
B	5.0	4.6
B High solid	4.8	5.2

Dynosol and Aerophen are both used at nominally 45% solids unlike
"A" and "B" which are used at 43% hence the final result comprises the
effect of increasing resin solid content of "B" which appears to be
beneficial when gluing Afara under standard mill conditions.

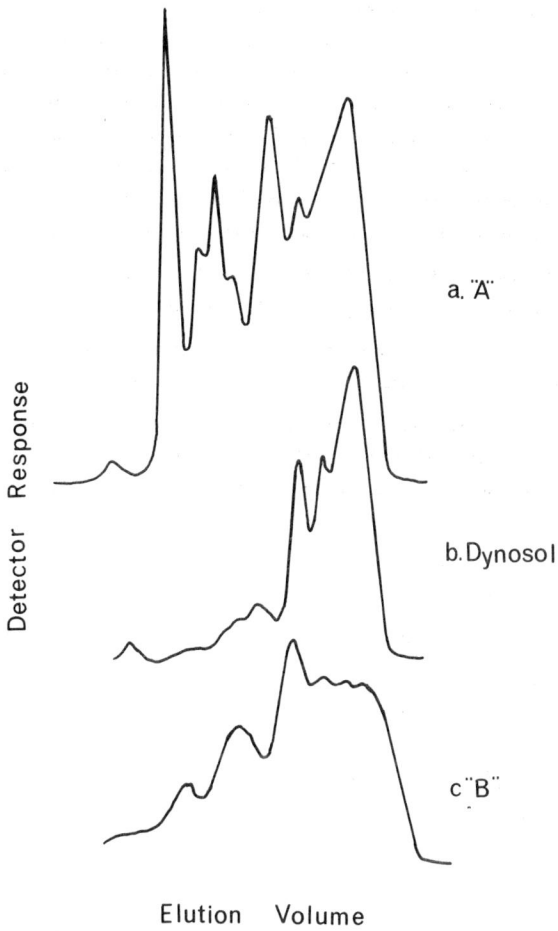

Fig. III   HPGPC Chromatograms of Phenolic Resins

3 THE BEHAVIOUR OF STERCULIA IN BONDING TRIALS

White Sterculia is a hard, yellow orange timber with numerous high orange rays giving the timber a very characteristic, attractive surface. The veneers have a very distinctive, strong odour which is even more marked after hot pressing. In bonding trials, the BS 1455 WBP knife values are always the lowest when bonding Sterculia to Sterculia. This makes sterculia about the most diffcult veneer to glue satisfactorily. Examination of Sterculia to Sterculia experimental bonds showed that:

(i)   The glue tended to 'dry out' very quickly when spread on the Sterculia veneers and did not penetrate the veneer.

(ii)   The bond always broke leaving most of the glue on the core i.e. there was little glue transference during hot pressing.

This behaviour indicates the gluing problem may be associated with a non-wetting of the surface which is defined in Young's concept of the contact angle formed by a drop of liquid resting on a plane solid surface as shown in Fig. IV.

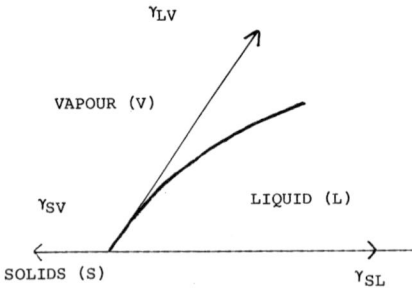

FIG. IV

If the drop does not spread then it is assumed to be held in equilibrium by the surface energies γ which are expressed mathematically as vector quantities in the Young equation

$$\gamma_{SV} = \gamma_{SL} + \gamma_{LV} \cos \theta \qquad\qquad \text{eq. 1}$$

where the subscripts SV, LV refer to the solid and liquid in equilibrium with the vapour respectively and SL the solid-liquid interface. When θ > 90° the liquid is non-spreading and when θ = 0° the liquid is said to wet the surface, the rate of wetting depending on the viscosity of the liquid and the roughness of the surface.

The work of Wenzel, ref. 12, suggests that the wetting properties of a solid should be directly proportional to the surface roughness, i.e. there is always an entrapment of air in the 'valleys' and the adhesive only wets the 'peak', and he defined a surface roughness factor r given by eq. (2)

$$r = \frac{\text{Total surface area}}{\text{Apparent geometric area}} \qquad \text{eq. 2}$$

which is used to modify θ by eq. 1:

$$\cos \theta' = r \cos \theta \qquad \text{eq. 3}$$

Contact angles were measured using drops of pure water on 3 veneers, namely Sterculia, Lagoswood and Afara with a Rane-Hart instrument to find if we were in a wetting or non-wetting situation. The results, given in Table IV show that the contact angles unfortunately decreased quite rapidly with time.

Table IV

Variation of contact angle with time

Timber Species	Contact Angles					
	0	2 min	4 min	6 min	8 min	10 min
Sterculia	126°	101°	89°	81°	71°	61°
Lagoswood	122°	75°	65°	61°	58°	-
Afara	31°	18°	13°	-	-	-

From the table the order of wetting hence the indicated order of cohesion is:

Afara > Lagoswood > Sterculia

The enhanced diffusion of glue into Afara which leads to poor adhesion is indicated by the very low contact angle found with this species and the very high contact angle in Sterculia indicates surface roughness must be contributing to this value.

Dougal, ref. 13, et al suggest a novel approach to solve the problem by removal of the veneer surface by sanding (or planing). Fig. V (a), (b) and (c) shows scanning electron micrographs of Sterculia,

Sanded

FIG. V (a)   Scanning Electron Micrographs of Sterculia
              Magnification   x50

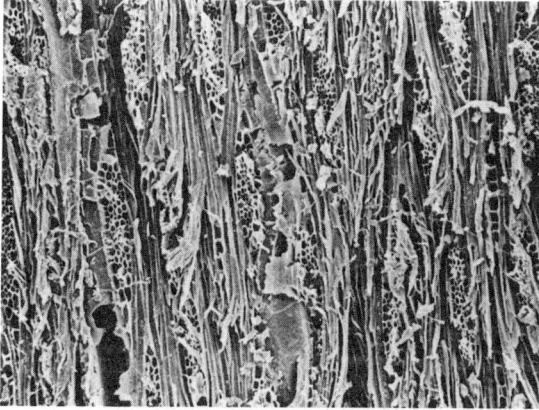

Sanded

FIG. V (b)   Scanning Electron Micrographs of Lagoswood

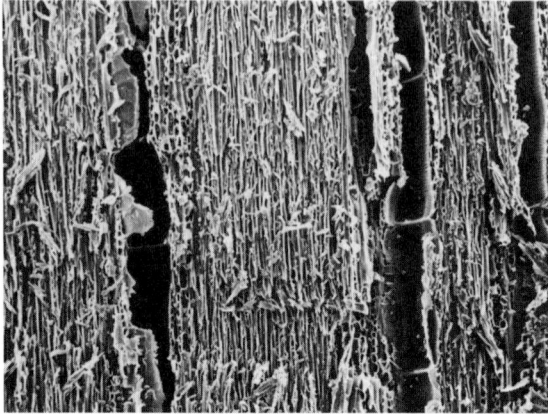

Sanded

FIG. V (c)   Scanning Electron Micrographs of Afara

Lagoswood and Afara veneers before and after a sanding treatment which does show the obvious smoothing effect, the radial parenchyma being completely filled with sanding debris.

The contact angles for the three timbers were measured after sanding to find if this treatment affected the contact angle and hence wettability, the results are shown in Table V.

Table V

Effect of sanding and contact angle

Timber Species	Initial Contact Angle	
	Before Sanding	After Sanding
Sterculia	126°	95°
Lagoswood	122°	51°
Afara	31°	53°

Both Sterculia and Afara agree with Wenzel's theory i.e. both approach 90° Lagoswood however is odd as it falls below 90°.

To demonstrate the effect of sanding on adhesion 5 ply Sterculia, Celtis, Afara, Pterygota (pterygota bequaertii) and Lagoswood boards were prepared 0.01" being sanded off each side prior to bonding. The knifing results of the sanded boards and unsanded controls using resin "A" are shown in Table VI.

Table VI

The effect of sanding on different species of timber veneers

Treatment		Dry Knife	BS1455:WBP 12 hr steaming
Sterculia	None	2.0.0.2	4.0.0.2
	Sanded	6.2.3.5	5.2.4.4
Celtis	None	4.0.0.5	5.3.0.4
	Sanded	8.4.2.6	6.4.2.5
Afara	None	6.6.4.5	5.4.2.3
	Sanded	7.6.6.7	7.4.4.4
Pterygota	None	6.5.5.6	5.3.2.4
	Sanded	7.6.8.6	7.4.4.3
Lagoswood	None	4.2.4.3	2.2.4.4
	Sanded	8.4.8.8	3.2.2.6

In every case better bonding was obtained after sanding
particularly for Sterculia and Celtis where delamination occurred in some
cases after steaming for the unsanded controls.

Although this seems to be a viable empirical explanation for the
improved bonding observed on sanding it was decided to carry out a further
test to verify if this effect was the only mechanism, or if chemical
compounds in the veneer surface contributed to the non-bonding mechanism.

As wood consists of approx 60% cellulose and approx 35% lignin with
only 5% or less of other chemicals the problem of identifying any material
on the surface which could hinder adhesion or affect the resin is very
difficult. Infra red analysis of Sterculia, Lagoswood and Afara showed
mainly lignin, cellulose and some minor bands indicating fatty materials
present, particularly in Sterculia.

Extraction of Afara veneer with boiling diethyl ether for 1 hour
gave no extract but extraction of Sterculia gave 0.66% w/w yield of a
brown oil. GLC analysis of the oil showed it to consist mainly of free
fatty acids and glycerides of chain lengths from C_{14} to C_{24} of which
approx 30% was either mono, di or triunsaturated.

Extraction of a sanded Sterculia veneer gave inexplicably a 1% w/w
yield of oil, i.e. almost twice as much as on the unsanded sample.

From these observations on the surface of Sterculia it would
initially appear that the presence of high levels of extractables
contributes to the poor adhesion, however more extractible oil is present
on the sanded veneer which in turn gives better adhesion. A possible
explanation could be that the unsanded, aged veneers contain insoluble
highly polymerised unsaturated oil on the surface, possibly arising during
the high temperature air drying to which the veneers are subjected, which
would not extract with ether. The presence of this polymerised material
could act as a type of 'paint' sealing the veneer surface and hence
preventing glue wetting and penetration. Sanding would of course remove
this film exposing a new surface which although containing more
extractibles is nevertheless more easily wetted and hence gives better
adhesion.

4 CONCLUSIONS

From this initial investigation into the differences in gluing effectively certain species of African hardwood, the following conclusions can be drawn:

1. When plywood boards containing wholly white Afara, Lagoswood or Sterculia are bonded under simulated standard plymill conditions using A resin a very poor bond is produced.

2. This poor bonding variability was demonstrated to be due to one of two effects:

    (a) Over penetration of the glue into the veneer leading to starved glue lines. This was observed with Afara and Lagoswood and is referred to as an "Afara type" difficult-to-glue problem.

    (b) Little or no penetration into the veneer which was observed for Sterculia ("Sterculia-type" difficult to glue).

3. For bonding Afara type species it was found that by producing a resin with a high proportion of high molecular weight material (resin B) and increasing the resin solids content by 2% this diffusion could be reduced to give acceptable bonding of Afara-type species to each other under simulated plymill conditions.

4. The bonding of Sterculia type difficult to glue species i.e. Sterculia and Celtis with phenolic resin is substantially improved by lightly sanding both sides of the veneer to remove 0.01" per side prior to gluing.

5. Other species examined, Afara, Pterygota and Lagoswood, also showed improved adhesion with phenolic resin after sanding.

6. Two reasons are proposed for this improvement in bonding on sanding. Contact angle measurements show that the removal of the surface roughness is a contributory factor and chemical analysis indirectly indicates the possibility of polymerised unsaturated oil on the surface which may be preventing adhesion by sealing the veneer surface.

REFERENCES

1. Adhesion and Adhesives: R Houwink, G Salmon, Vol. I, p1-128, 1970.

2. Hariaka, K Proceedings IUFRO Div. 5, Sept 1973, 2, 503-27.

3. Okano, A, Bull Soc. Govt Forest Exp Section Japan, 230, 143-54 (1970).

4. Troughton, G E, Chow, S, J Inst. Wood Science, 21, 120-25 (1968).

5. Klein, Chemical Aspects of Gluing Plywood with Phenolic Resins,
      FRPS Separate No. AM-75-567.

6. Duval, M, Block, B, Kohn, S. J. Appl. Pol. Sci. 16, 1585-1602 (1972).

7. Sebenik, A. J. Chromatography 160, 205-212 (1978).

8. Wagner, E R, Greff, R J. J. Pol. Sci. Al. 9, 2193-2207 (1971).

9. Cazes, J, Martin, N. in Liquid Chromatography of Polymers and Related
      Materials. Ed. J Cazes Chromatographic Science Series Vol. 8.
      (Dekker, 1976).

10. Matsuzaki, T. et al, J. Liquid Chromatog. 3, 3, 353-365 (1980).

11. Pizzi, A, Scharfetter, H O. J. Appl. Pol. Sci. 22 1745-1761 (1978).

12. Wenzel R.A. Ind. Eng. Chem. 28, 988, 1936.

13. Dougal, E.F. Glue-line Characteristics and Bond Durabilty of Southwest
      Asian Species after Solvent Extraction and Planing of
      Veneer, Forest Prod. J. 30 (7) 48-53 (1980).